D0204603

MICHIGAN MOLECULAR INSTITUTE
1910 WEST ST. ANDREWS ROAD
MIDLAND, MICHIGAN 48640

Superabsorbent Polymers

ACS SYMPOSIUM SERIES **573**

Superabsorbent Polymers

Science and Technology

Fredric L. Buchholz, EDITOR

Dow Chemical Company

Nicholas A. Peppas, EDITOR

Purdue University

MICHIGAN MOLECULAR INSTITUTE
1910 WEST ST. ANDREWS ROAD
MIDLAND, MICHIGAN 48640

Developed from a symposium sponsored
by the Division of Polymeric Materials: Science and Engineering, Inc.
at the 206th National Meeting
of the American Chemical Society,
Chicago, Illinois,
August 22–27, 1993

American Chemical Society, Washington, DC 1994

Library of Congress Cataloging-in-Publication Data

Superabsorbent polymers: science and technology / Fredric L. Buchholz, editor, Nicholas A. Peppas, editor.

  p.   cm.—(ACS symposium series, ISSN 0097–6156; 573)

"Developed from a symposium sponsored by the Division of Polymeric Materials: Science and Engineering at the 206th National Meeting of the American Chemical Society, Chicago, Illinois, August 22–27, 1993."

Includes bibliographical references and indexes.

ISBN 0–8412–3039–0

1. Acrylic resins—Congresses. 2. Sorbents—Congresses.

I. Buchholz, Fredric L., 1952–   . II. Peppas, Nicholas A., 1948–   . III. American Chemical Society. Division of Polymeric Materials: Science and Engineering. IV. American Chemical Society. Meeting (206th: 1993: Chicago, Ill.) V. Series.

TP1180.A35S76   1994
668.4′232—dc20   94–34669
                                                                    CIP

The paper used in this publication meets the minimum requirements of American National Standard for Information Sciences—Permanence of Paper for Printed Library Materials, ANSI Z39.48–1984.   ∞

Copyright © 1994

American Chemical Society

All Rights Reserved. The appearance of the code at the bottom of the first page of each chapter in this volume indicates the copyright owner's consent that reprographic copies of the chapter may be made for personal or internal use or for the personal or internal use of specific clients. This consent is given on the condition, however, that the copier pay the stated per-copy fee through the Copyright Clearance Center, Inc., 27 Congress Street, Salem, MA 01970, for copying beyond that permitted by Sections 107 or 108 of the U.S. Copyright Law. This consent does not extend to copying or transmission by any means—graphic or electronic—for any other purpose, such as for general distribution, for advertising or promotional purposes, for creating a new collective work, for resale, or for information storage and retrieval systems. The copying fee for each chapter is indicated in the code at the bottom of the first page of the chapter.

The citation of trade names and/or names of manufacturers in this publication is not to be construed as an endorsement or as approval by ACS of the commercial products or services referenced herein; nor should the mere reference herein to any drawing, specification, chemical process, or other data be regarded as a license or as a conveyance of any right or permission to the holder, reader, or any other person or corporation, to manufacture, reproduce, use, or sell any patented invention or copyrighted work that may in any way be related thereto. Registered names, trademarks, etc., used in this publication, even without specific indication thereof, are not to be considered unprotected by law.

PRINTED IN THE UNITED STATES OF AMERICA

1994 Advisory Board

ACS Symposium Series

M. Joan Comstock, *Series Editor*

Robert J. Alaimo
Procter & Gamble Pharmaceuticals

Mark Arnold
University of Iowa

David Baker
University of Tennessee

Arindam Bose
Pfizer Central Research

Robert F. Brady, Jr.
Naval Research Laboratory

Margaret A. Cavanaugh
National Science Foundation

Arthur B. Ellis
University of Wisconsin at Madison

Dennis W. Hess
Lehigh University

Hiroshi Ito
IBM Almaden Research Center

Madeleine M. Joullie
University of Pennsylvania

Lawrence P. Klemann
Nabisco Foods Group

Gretchen S. Kohl
Dow-Corning Corporation

Bonnie Lawlor
Institute for Scientific Information

Douglas R. Lloyd
The University of Texas at Austin

Cynthia A. Maryanoff
R. W. Johnson Pharmaceutical
    Research Institute

Julius J. Menn
Western Cotton Research Laboratory,
    U.S. Department of Agriculture

Roger A. Minear
University of Illinois
    at Urbana–Champaign

Vincent Pecoraro
University of Michigan

Marshall Phillips
Delmont Laboratories

George W. Roberts
North Carolina State University

A. Truman Schwartz
Macalaster College

John R. Shapley
University of Illinois
    at Urbana–Champaign

L. Somasundaram
DuPont

Michael D. Taylor
Parke-Davis Pharmaceutical Research

Peter Willett
University of Sheffield (England)

Foreword

THE ACS SYMPOSIUM SERIES was first published in 1974 to provide a mechanism for publishing symposia quickly in book form. The purpose of this series is to publish comprehensive books developed from symposia, which are usually "snapshots in time" of the current research being done on a topic, plus some review material on the topic. For this reason, it is necessary that the papers be published as quickly as possible.

Before a symposium-based book is put under contract, the proposed table of contents is reviewed for appropriateness to the topic and for comprehensiveness of the collection. Some papers are excluded at this point, and others are added to round out the scope of the volume. In addition, a draft of each paper is peer-reviewed prior to final acceptance or rejection. This anonymous review process is supervised by the organizer(s) of the symposium, who become the editor(s) of the book. The authors then revise their papers according to the recommendations of both the reviewers and the editors, prepare camera-ready copy, and submit the final papers to the editors, who check that all necessary revisions have been made.

As a rule, only original research papers and original review papers are included in the volumes. Verbatim reproductions of previously published papers are not accepted.

M. Joan Comstock
Series Editor

Contents

INDEXES

Preface

SUPERABSORBENT POLYMERS ARE USED TO ABSORB body fluids in a variety of personal care products, notably baby diapers. These polymers absorb and hold under pressure about 30 times their weight in urine and help keep the baby's skin much drier and healthier. Because of the effectiveness of superabsorbent polymers, diapers have become thinner as polymer replaces the bulkier cellulose fluff. The industry has grown from an idea in the early 1970s to commercial viability in Japan in the early 1980s and in the United States in the mid-1980s to nearly a one billion pound per year global industry today. The growth of the superabsorbent polymer industry and the competitive drive toward new, thinner, and less costly diaper designs requires polymers with improved properties. A flurry of competitive research effort resulted in several different approaches toward meeting the changing needs of the diaper industry. As a result, techniques for making such polymers have advanced significantly in the past ten years.

For this book, as for the symposium on which the book was based, we wanted to find a balance between the applied polymer science of industry and the fundamental polymer science available at universities. The global nature of the superabsorbent polymer industry meant getting perspectives of the technology from around the world. We brought together industrial and academic experts from Japan, North America, and Europe to cover topics as diverse as the type of polymers needed for personal care, new theoretical descriptions of polyelectrolyte networks, and emerging applications for superabsorbent polymers. We hope you find the book to be a useful description of the technology of superabsorbent polymers.

FREDRIC L. BUCHHOLZ
Specialty Chemicals Research and Development Department
1603 Building
Dow Chemical Company
Midland, MI 48674

NICHOLAS A. PEPPAS
School of Chemical Engineering
Purdue University
West Lafayette, IN 47907–1283

March 25, 1994

SYNTHESIS

Chapter 1

Formation and Structure
of Cross-Linked Polyacrylates
Methods for Modeling Network Formation

A. B. Kinney and A. B. Scranton

Department of Chemical Engineering, Michigan State University,
East Lansing, MI 48824

Models developed to describe free radical crosslinking polymerizations may be divided into three classes: statistical models, kinetic descriptions, and kinetic gelation simulations. This paper provides a comparison of the three types of models by describing the approach, assumptions, strengths, and limitations of each when applied to lightly crosslinked systems. The greatest strength of the statistical approach arises from the wealth of post-gel structural information that can be obtained, however these models have difficulty accounting for history-dependent effects. Kinetic descriptions may properly describe history-dependent reaction non-idealities, but are unable to account for localized effects that are prevalent in highly crosslinked systems. The kinetic gelation simulations may describe topological constraints and localized effects but are of limited use for lightly crosslinked polymers due to their inability to accurately account for molecular motions.

In recent years there has been considerable interest in water-swellable "super-absorbing" polymers capable of absorbing and holding large amounts of water. These polymers have found extensive commercial application as sorbents in personal care products such as infant diapers, feminine hygiene products, and incontinence products (1,2), and have received considerable attention for a variety of more specialized applications including matrices for enzyme immobilization (3), biosorbents in preparative chromatography (3), materials for agricultural mulches (4), and matrices for controlled release devices (5). The interest in superabsorbing polymers is illustrated by the popularity of the ACS symposium upon which this volume is based, as well as the recent proliferation of research articles on the topic (for reviews, see references 1-7).

0097–6156/94/0573–0002$08.90/0
© 1994 American Chemical Society

Absorbent polymers are most commonly formed by free radical crosslinking polymerizations of hydrophilic acrylate or methacrylate monomers with small quantities of crosslinking agents containing two (or more) polymerizable double bonds (2-10). Examples of typical crosslinking agents include *N,N'*-methylenebisacrylamide, triallylamine, ethyleneglycoldiacrylate, tetraethyleneglycoldiacrylate, trimethylolpropanetriacrylate, and the methacrylate analogs of the aforementioned acrylates. Hydrophilic esters of acrylic or methacrylic acid (such as 2-hydroxyethylmethacrylate and its analogs) have been extensively polymerized using these reactions to form hydrogels which typically exhibit a maximum swelling of 40-50 wt% water. The unique set of properties exhibited by the hydroxyethyl(meth)acrylate hydrogels have lead to their extensive application in a host of biomedical materials and devices (*11 - 13*). Hydrogels exhibiting much higher swelling capacities may be produced by carrying out the free radical crosslinking polymerizations with ionogenic monomers such as acrylic and methacrylic acid (or their sodium salts). For example, poly(acrylic acid) hydrogels may exhibit a maximum swelling of more than 99 wt% water.

The editors of this volume have asked us to provide a tutorial on methods for modeling network formation in free radical crosslinking polymerizations of hydrophilic acrylates and methacrylates. Models which have been developed to describe free radical crosslinking polymerizations may be divided into three classes: i) statistical models, ii) kinetic descriptions, and iii) kinetic gelation models based upon computer simulated random walks. This tutorial will provide a comparison of the three types of models by describing the approach, assumptions, strengths, and limitations of each when applied to lightly crosslinked systems. Because each of the three approaches could serve as the basis of a lengthy review, our discussion will be representative rather than exhaustive. Similarly, related topics such as preparative methods for superabsorbing polymers (*2,14 ,15*) and kinetic anomalies exhibited by ionogenic monomers (*16 -20*) have been reviewed in the past and will not be extensively discussed in this contribution.

Characteristics of Free Radical Crosslinking Polymerizations

When a monomer containing one polymerizable double bond (such as an acrylate) is copolymerized with small amounts of a divinyl crosslinking agent, simultaneous copolymerization and crosslinking reactions take place. For example, a growing radical may react with one of the double bonds of a crosslinking agent to form a polymer chain with a pendant double bond. This pendant double bond may react with a second growing chain to form a crosslink between chains. It is the presence of the crosslinks between chains that leads to the most notable characteristic of these polymerizations: the possible formation of a three-dimensional polymer network. At a conversion called the gel point, enough chains are linked together to form a large macromolecular network that spans the entire reaction system (its size is limited only by the size of the system). The gel point is of considerable practical importance because the characteristics of the reaction system change considerably: the system undergoes a transition from a viscous liquid to an elastic solid (the viscosity diverges), the weight-average molecular weight diverges, and the first formation of an insoluble gel phase occurs. It is worth noting that despite the large change in physical properties at the gel point, the initial gel molecule (the first spanning structure) is only a very small fraction of the reaction system (in fact, the gel weight fraction is nearly zero). After the gel point, the gel fraction grows as monomer units and polymer chains are rapidly incorporated into the network structure.

Other characteristics of free radical crosslinking copolymerizations have been elucidated by many authors. Several investigations have demonstrated the importance of intramolecular cyclization, which occurs when a growing polymer chain reacts with a pendant double bond to which it is already connected. Intramolecular cyclization is important because it results in the consumption of a pendant double bond without contributing to the polymer network structure. In free radical crosslinking polymerizations, evidence of cyclization has been obtained by measurements of the gel point conversion (*21 -24*) and pendant double bond conversion (*25 -27*). The probability of cyclization also depends on the chain length between double bonds of the crosslinking agent (*28 -30*). If the distance between double bonds is very small or very large, the probability of cyclization was found to be relatively small. However, a maximum in the probability of cyclization was found to occur when the dimethacrylate crosslinking agent contained approximately nine oxyethylene units.

If a free radical crosslinking polymerization is carried out using relatively high concentrations of crosslinking agent, the extent of cyclization may become so high that the reaction mixture becomes heterogeneous, with small, isolated micro-gel regions of high crosslink density (*31 -34*). The first indirect experimental evidence of the formation of microgels was provided by intrinsic viscosity measurements (*35*) that were more characteristic of a suspension of small gel particles than a solution of randomly branched polymers. More recently, Spevacek and Dusek (*36*) studied the integrated peak intensities in ^{13}C and ^1H NMR spectra for low-conversion copolymers of styrene with divinyl benzene. These authors found that while all of the repeating units showed up in the carbon spectra, a significant fraction were absent from the proton spectra. They attributed this discrepancy to the presence of microgels in which the mobility of the repeating units was low, thereby causing the proton relaxation to occur too rapidly to be detected. More recently, evidence for the presence of microgels has been based upon fluorescence polarization (*33*), electron paramagnetic resonance (*37*), and NMR relaxation (*38*) studies.

In free radical crosslinking polymerizations, the concentration of the crosslinking agent may have a significant effect on the observed magnitude of the gel effect (autoacceleration). Autoacceleration is a manifestation of the diffusion-controlled nature of the termination reaction, and is characterized by a decrease in the rate of termination which leads to an increase in the polymerization rate (*39*). In free radical crosslinking polymerizations, the mobility of the growing chains depends upon the number of crosslinks they have formed with other chains, therefore the magnitude of the gel effect increases as the content of the crosslinking agent is increased. For example, Scranton *et al.* (*40*) used differential scanning calorimetry to study copolymerizations of ethylene glycol monomethacrylates with small amounts of dimethacrylate crosslinking agents. In these studies, the time duration before the onset of the gel effect was found to decrease and the maximum rate was found to increase as the crosslinker concentration was increased from 0 to 2 mol%. Similar trends were observed by Ulbrich *et al.* (*41 ,42*) for reactions of butyl and ethyl methacrylamide with methylene bisacrylamide.

A phenomenon related to the gel effect is an increase in the radical concentration during the course of the reaction. If the rate of termination decreases during the course of the reaction, it is reasonable to expect that the radical concentration will increase. However, until recently no direct measurement of the radical concentration during a free radical polymerization had been reported because the electron spin resonance (ESR) spectrometers lacked sufficient sensitivity (*43*). Recently, Shen and Tian, (*44*) used an ESR spectrometer with enhanced sensitivity to measure the *in situ* radical concentration during a bulk polymerization of methyl methacrylate. Zhu *et al.* (*45*) applied the technique to free radical crosslinking reactions when they measured the radical concentration as a function of time for the copolymerization of methyl methacrylate with 0.3 to 100 wt% ethylene glycol dimethacrylate (EGDMA) cross-linking agent. For EGDMA concentrations lower than one percent, the radical concentration profile initially remained fairly constant, and the quasi-steady state hypothesis appeared to be valid (*45*). Eventually, however, the radical concentration increased dramatically to a level more than an order of magnitude higher than the initial steady-state value. This sudden increase in the radical concentration coincided with an increase in the reaction rate. Subsequently, the radical concentration exhibited a small drop, followed by a gradual increase. The time at which this sudden rise occurred was found to decrease as the concentration of EGDMA was increased. For higher EGDMA concentrations, the quasi-steady state assumption was invalid throughout the reaction.

Modeling of the Formation of Crosslinked Polymers

As stated previously, models which have been developed to describe free radical crosslinking reactions may be divided into three classes: i) statistical models, ii) kinetic descriptions, and iii) kinetic gelation models based upon computer simulated random walks. Each type of model will now be considered separately with emphasis on their advantages and limitations when applied to lightly crosslinked systems.

Statistical Models. Statistical models consider an ensemble of polymer structures built from monomer units according to (mean field) probabilistic rules for bond formation. The recursive nature of the polymer structure is exploited to obtain expressions for statistical averages, including network structural information. Several statistical computational schemes have been reported in the literature, including the recursive method, (*46 ,47*) the fragment method (*48*), and the generating function method (*49 ,50*). Several equivalent statistical descriptions of free radical crosslinking polymerizations are based upon the assumptions of ideal random crosslinking, including: i) equal and independent reactivity of all carbon double bonds, ii) no intramolecular cyclization, iii) termination by chain transfer or disproportionation, and iv) conversion independent kinetics. The approach has recently been extended to include conversion dependent kinetics (*46,51*), multiple termination mechanisms (*46,48,50*) and approximate treatment of cyclization (*47,49*).

In our discussion of the statistical approach for modeling free radical crosslinking polymerizations, we will concentrate on the generating function approach for developing the expressions for structural averages. As mentioned above, several equivalent computational schemes have been reported in the literature, and the reader is referred to references 46-52 for details on the alternative methods. In all statistical models structural averages are taken over an ensemble of polymer chains that have been constructed according to a set of probabilistic rules for bond formation. In the generating function method, the rules for bond formation are described by a set of probability generating functions for the number of bonds issuing from the monomer units. As shown below, a generating function, F(s), simply represents a sum over all possible bonded states of the statistical weight for each state multiplied by a product of dummy variables that identify the state.

$$F(\mathbf{s}) = \sum_i \sum_j P_{ij} s_1^i s_2^j \tag{1}$$

Here P_{ij} is the statistical weight for a unit which has i number of bonds with other units of type 1 and j bonds to units of type two. Once the generating functions have been formulated, expressions for a variety of structural averages may be obtained in a straight-forward manner using the mathematical framework of the theory of branching processes (52 -55).

Formulation of the Link Probability Generating Functions. Before discussing the formulation of the generating functions for our reaction system, a descriptive picture of the generating function technique may be useful. For pedagogical purposes, the mathematical treatment can be envisioned in the following manner: the polymerization is allowed to proceed to the conversion of interest, then is suspended in time so that the ensemble of polymer chains that have formed can be analyzed. A monomer unit (acrylate or crosslinker) is chosen at random (each unit with the same probability) and is labeled as being in the zeroth topological generation. Any monomer unit directly bonded to this unit is in the first topological generation, and so on. The link probability generating function, $\mathbf{F}_0(\mathbf{s})$ describes the distribution of bonds between the zeroth and first generations, and the generating function, $\mathbf{F}(\mathbf{s})$, describes this distribution for all other generations.

For free radical crosslinking polymerizations, the generating functions $\mathbf{F}_0(\mathbf{s})$ and $\mathbf{F}(\mathbf{s})$ are two-component vectors with an element for each type of monomer unit (49):

$$\mathbf{F}_0(\mathbf{s}) = [F_{01}(\mathbf{s}), F_{02}(\mathbf{s})]^T \tag{2}$$

$$\mathbf{F}(\mathbf{s}) = \left[F_1(\mathbf{s}), F_2(\mathbf{s}) \right]^T \tag{3}$$

Likewise, the dummy variable vector, s also has a component for each type of monomer:

$$\mathbf{s} = \left[s_1, s_2 \right]^T \tag{4}$$

Here, the subscript 0 refers to the zeroth generation while the subscripts 1 and 2 refer to the acrylate and the crosslinking agent respectively. Note that each vector contains two components: the first corresponding to the acrylate monomer and the second corresponding to the doubly unsaturated crosslinking agent. In this paper, vectors are denoted by bold print, while the superscript T denotes transpose.

To illustrate the formulation of link probability generating functions for a free radical crosslinking polymerization, six possible bonded states of the acrylate monomer in a free radical crosslinking polymerization are illustrated in Figure 1. Note that the bonded states of a monomer unit are described by two attributes: 1) the number of bonds the unit has formed with other monomers and 2) the type of unit(s) to which it is bonded. Therefore, as illustrated in the figure, the monovinyl acrylate monomer has the following six possible bonded states: unreacted, therefore no bonds with other units (state 1); reacted to form one bond with another unit, which may be another acrylate monomer (state 2, identified by dummy variable s_1 in the generating function); or a crosslinker (state 3, denoted by dummy variable s_2 in the generating function); reacted to form two bonds with other units, which may be two other acrylate monomers (state 4, identified by dummy variable product s_1^2); an acrylate and a crosslinker (state 5, denoted by $s_1 s_2$ in the generating function); or two crosslinkers (state 6, denoted by s_2^2). A similar analysis yields fifteen possible reacted states for the doubly unsaturated crosslinking agent.

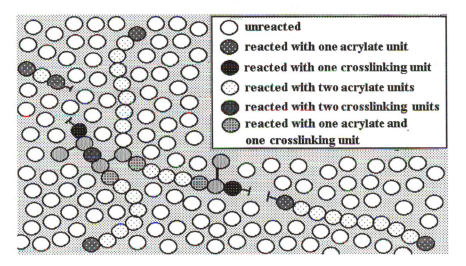

Figure 1. Monomer reacted states of the acrylate monomer during a free radical crosslinking polymerization.

With this set of possible bonded states in mind, an illustrative set of link probability generating functions for free radical crosslinking polymerizations are shown below (*49*). This simplest set of generating functions is based upon the aforementioned assumptions of ideal random crosslinking. These assumptions can be relaxed within the statistical framework, as discussed later in this tutorial.

$$F_{01}(s) = 1 - a + a\{(1-p) + p[(1-r)s_1 + rs_2]\}^2 \tag{5}$$

$$F_{02}(s) = F_{01}(s) \tag{6}$$

$$F_1(s) = 1 - p + p[(1-r)s_1 + rs_2] \tag{7}$$

$$F_2(s) = \{1 - p + p[(1-r)s_1 + rs_2]\}(F_{01}) \tag{8}$$

In these equations (49), "a" corresponds the conversion of double bonds in the system; "p" is the probability that a growing radical will add at least one more monomer unit before terminating (equal to the rate of propagation divided by the sum of the rates of propagation, termination, and chain transfer); "r" is the fraction of the initial double bonds belonging to the crosslinker units; s_1 is the dummy variable for a bond with an acrylate unit, and s_2 is the dummy variable for a bond with a crosslinker unit.

The generating functions shown above include contributions for every possible bonded state of each monomer while accounting for the free-radical mechanism by which these bonds are formed (49). For the zeroth generation, the "1-a" term accounts for the double bonds which have not reacted, while "a" is the probability that a double bond has reacted. The bracket which multiplies "a" describes the possible bonded states upon reaction. The "1-p" term describes reacted functionalities which have no bonds to the next generation (this corresponds to the chain ends), while "p" is the probability that a reacted functionality has a bond issuing to the next generation. The coefficients of the dummy variables s_1 and s_2 are the probabilities the bond has formed with an acrylate unit and a crosslinking unit, respectively. Monomer units in all generations higher than the zeroth already have a bond to the previous generation, and will have a bond to the next generation with probability p. The fact that the crosslinker has twice as many double bonds as the acrylate monomer is accounted for by the higher order of the crosslinker generating functions.

Calculation of Structural Averages. The utility of the link probability generating functions $F_0(s)$ and $F(s)$ lies in the ease with which they may be used to obtain statistical averages over the ensemble of polymer chains built from the monomer units. Equations for many pre-gel and post-gel structural averages may be derived using these generating functions (55-58), including the pre-gel number and weight average molecular weights, and the post-gel sol weight fraction, molecular weight between crosslinks, and number of elastically active network chains. For example, an expression for the pre-gel weight average molecular weight may be readily derived by first producing the weight fraction generating function by cascade substitution of the link generating functions, then differentiating with respect to the dummy variables. This procedure is described in detail in reference 56; the final result is a closed-form, analytical expression for the weight average molecular weight, M_w, in terms the generating functions:

$$M_w = M^T(\Delta + (\Delta - \Gamma)^{-1}\Gamma_0)m \tag{9}$$

Here Γ_0 is the transposed Jacobian matrix of the zeroth generation probability generating function evaluated at $s=1$, Γ is the corresponding matrix for all other generations, Δ is the unitary matrix, M is the molecular weight vector, and m is the mass fraction vector.

$$m = [m_1, m_2]^T \tag{10}$$

$$M = [M_1, M_2]^T \tag{11}$$

The gel point is a quantity of considerable practical importance in free radical crosslinking polymerizations. Because the weight average molecular weight diverges at the gel point (*39,46-55*), an expression for the gel point conversion may be readily obtained from the above equation for the weight average molecular weight. At the gel point, the following equation holds:

$$|\Delta - \Gamma| = 0 \tag{12}$$

Substituting the generating functions (equations 5-8) into this equation yields the following expression for the gel point conversion:

$$a_c = \frac{1-p}{2pr} \tag{13}$$

One of the strengths of the statistical approach arises from the utility of the extinction probability vector, v which allows the sol fraction and the gel fraction (*49,59,60*) to be distinguished from one another, and more importantly, allows internal network structural averages to be determined. For example, the molecular weight of the sol fraction alone (ignoring the gel fraction) may be calculated (*53*) as well as gel characteristics such as the molecular weight between crosslinks (*49*), and the number of elastically active chains (*55*). Conceptually, the extinction probability is the probability that a bond to the next generation has no paths which lead to the infinite network. This statement implies that a bond in the current generation has no paths with infinite continuation only if all bonds in the next generation have no paths with infinite continuation. Therefore the extinction probability vector is given by the following set of simultaneous algebraic equations (*52,55*):

$$v = F(v) \tag{14}$$

Here v is the extinction probability vector (with one component for each type of monomer) and $F(v)$ is the probability generating function vector for all generations higher than the zeroth evaluated with $s = v$.

The extinction probability may be used to restrict attention to either the gel fraction or the sol fraction when calculating structural averages. If it is desired to consider only the sol fraction, all bonds are weighted by the probability that they do not issue a path with infinite continuation (in the generating functions, s is replaced by $\mathbf{v}$; i.e. s_1 is replaced by v_1 and s_2 is replaced by v_2). For example, the sol weight fraction, w_s may be calculated using the following equation (58):

$$w_s = m_1 F_{01}(\mathbf{v}) + m_2 F_{02}(\mathbf{v})$$ (15)

In a similar manner, expressions for other sol fraction structural averages have been developed and reported (53).

For calculation of network structural averages, it is useful to define another generating function, $\mathbf{T(s)}$, which describes the distribution of monomer units with bonds leading to the gel (50,56). To obtain this auxiliary generating function, each bond in the zeroth generation link probability generating function is weighted by the probability that the bond eventually leads to the gel (49,55):

$$\mathbf{T(s)} = \mathbf{F_0}(\mathbf{v}+(1-\mathbf{v})\mathbf{s}) = [T_1(s), T_2(s)]$$ (16)

Each of the components generating functions $T_1(s)$ and $T_2(s)$ represents a sum of the following form (49,55):

$$T_1 = \sum_i \sum_j t_{1(i+j)} s_1^i s_2^j$$ (17)

$$T_2 = \sum_i \sum_j t_{2(i+j)} s_1^i s_2^j$$ (18)

Here the subscripts 1 and 2 again refer to the acrylate monomer and the crosslinking agent, respectively. The index i represents the number of bonds between the current unit and other acrylate units that lead to the gel, while the index j represents the number of bonds with crosslinkers units that lead to the gel. The coefficients t_k, where $k = i + j$, represent the fraction of the units that issue k paths to the gel.

The internal structure of an absorbent polymer gel plays an important role in determining the material properties. For example, the mechanical properties are determined by the number of elastically active network chains, and are therefore related by the crosslink density. An effective crosslink must issue at least three paths to the infinite polymer network. Since there is an effective crosslink at each end of an active chain, each unit with more than two paths to the network $(k > 2)$ contributes $1/2$ of an elastically active network chain (EANC). Therefore, the number of EANCs, N_e, in a two-component copolymerization is given by the following equation (49):

$$N_e = \frac{1}{2}(n_1 \sum_{k=3}^{\infty} k t_{1k} + n_2 \sum_{k=3}^{\infty} k t_{2k})$$ (19)

In the free radical crosslinking polymerizations, the acrylate (monomer 1) may issue at most two bonds, therefore the first sum in the above equation is zero. Also, since a doubly unsaturated crosslinker may issue at most four bonds, the equation simplifies to the following:

$$N_e = \frac{1}{2}(n_2 \sum_{k=3}^{4} k t_{2k})$$ (20)

This sum may be evaluated using the derivatives of $T_2(s)$ to yield the following expression for the number of EANC:

$$N_e = \frac{1}{2}[T_2(1) - T_2(0) - T'_2(0)]$$ (21)

The molecular weight between crosslinks, M_c may be defined as the average molecular weight between effective crosslinks, or equivalently, the average molecular weight of an elastically active network chain. With this definition, the effects of dangling chain ends are accounted for (they do not contribute to the molecular weight between crosslinks), and the assumption of a perfect network is not necessary (*49,61*). The average number of repeating units in an elastically active network chain may be calculated using the generating function $T(s)$ (*54*):

$$L_e = \frac{t_2 + \sum\limits_{k=3}^{\infty} k t_k}{N_e}$$ (22)

This equation may be derived by considering that every monomer unit which bears $k > 2$ bonds leading to the gel is part of k elastically active chains, while every unit with exactly two such bonds is a member of one active chain, and units with zero or one such bonds do not contribute to any active chains. The term t_2 and the sum in this equation may be evaluated in terms of the derivatives of $T(s)$ to yield the following equation for the molecular weight between crosslinks (*49*):

$$M_c = \frac{M\left[\frac{1}{2} n_1 T'_1(0) + n_2\left(T_2(1) - T_2(0) - \frac{1}{2} T'_2(0)\right)\right]}{N_e}$$ (23)

Here M is an average monomer molecular weight:

$$M = \frac{n_1 M_1 + n_2 M_2}{n_1 + n_2}$$ (24)

These two equations, when combined with the expressions for the probability generating functions, provide closed-form, analytical expressions for the molecular weight between crosslinks as a function of conversion (*49*).

Generalization of the Statistical Approach. The above discussion illustrates the general manner in which the statistical approach may be implemented, however the simple set of generating functions we have considered thus far is based upon the rather restrictive assumptions of ideal random crosslinking and a brief discussion on how these assumptions may be relaxed is in order. Due to the limited scope of this tutorial, we will provide only an overview of how these assumptions may be relaxed. For more details, the reader is referred to references (*46-60*).

If the assumption of equal and independent reactivity of all double bonds in the system is invalid, the statistical weights in the link probability generating functions must be modified (49). For example, if one monomer is more reactive than the other, then the compositional term "r" will not remain constant, but will change with conversion. Similarly, effects such as the gel effect and drift in monomer concentration may cause the probability of propagation, "p" to depend upon conversion (46,49). To account for these and other conversion-dependent kinetic effects, the statistical weights can be derived by solving the kinetic differential equations that describe the concentrations of the reactive species (62). This combined statistical and kinetic approach exploits the strengths of both methods: solution of the kinetic equations allows conversion-dependent kinetic effects to be accurately modeled in the formulation of the statistical weights, while the statistical analysis allows network structural information to be derived through the use of the extinction probability.

As discussed earlier in this paper, intramolecular cyclization occurs when a growing chain reacts with a pendant double bond to which it is already connected. The cyclization reaction is important because it results in the consumption of a pendant double bond without linking tow chains together. Therefore the presence of cyclization moves the gel point to higher conversion, decreases the number of effective crosslinks, and decreases the number of elastically active chains in the final network. Cyclization is very difficult to accurately account for in a statistical description (60), however cyclization may be accounted for in an approximate manner by effectively disregarding the bonds that form intramolecular cycles (47,49). Therefore, a cycle-forming functionality is considered to be in a reacted state, but without a bond extending to the next generation.

Multiple termination mechanisms may be accurately included into a statistical description of free radical crosslinking polymerizations by carefully accounting for the directionality of the free radical polymerization (46,48,50). If termination occurs in part by combination, the reaction directionality causes the direction of chain propagation to be statistically distinct from the opposite direction along the polymer chain. Therefore, a bond formed in the direction of propagation is not statistically equivalent to a bond formed opposite the direction of propagation, and, in the generating function approach, two dummy variables must be defined for each type of monomer unit (50). As a result, the zeroth generation link probability generating function, $F_0(s)$, is a two-component vector (one component for each type of monomer), however, the generating function for all higher generations, $F(s)$, is a four-component vector with two components for each type of monomer unit (one for each direction along the chain) (50). This difference in dimensionality requires some of the expressions for the structural averages to be modified slightly, however the general procedure for obtaining statistical averages is essentially the same as we have outlined for the simplest set of generating functions.

Advantages and Limitations of the Statistical Approach. Statistical models offer both advantages and limitations for description of lightly crosslinked polymers formed in free radical crosslinking polymerizations. Perhaps the greatest strength of the statistical approach arises from the wealth of network structural information that may be obtained.   The use of the extinction probability vector allows many kinetically-equivalent structural features be distinguished from one another.   For example, dangling chain ends may be distinguished from elastically active chains (*59*), the number of elastically active crosslinks units may be determined, etc. Therefore, a statistical analysis may to provide the type of structural information that is of prime interest when considering the mechanical, sorption, and permeability characteristics of the polymer network.

The statistical approach also has several disadvantages and limitations.   A considerable limitation for application to free radical crosslinking polymerizations is the fact that the statistical approach most naturally describes conversion-independent, equilibrium-controlled systems (*59,60,64*) since all of the polymer chains are generally treated as statistically equivalent.   Free radical polymerizations are actually kinetically controlled, therefore the structure of the polymer chains depends upon the conversion at which they were formed, and the history of the process is important (*59,60,64*).   For example, reaction conditions (concentrations, kinetic constants, etc.) may change considerably during the course of the reaction.   This time dependency may affect such quantities as the instantaneous primary chain length, and the probability of reaction with a pendant double bond.   Therefore, polymer chains formed at low conversions may have considerably different structure than those formed at high conversions.   Although such history-dependent effects may be included in a statistical description (*65*), the analysis is considerably more complicated than the procedure outlined here.

Other limitations of the statistical approach arise from the mean field approximation upon which it is based (*66*).   The probabilistic rules for bond formation used to formulate the expressions for the structural averages are invariably derived using mean field values for concentrations, kinetic constants, etc.   While the mean field approach is typically valid for homogeneous reaction systems in which all molecules of a given type are equivalent, it is not valid for free radical crosslinking polymerizations containing a preponderance of tetrafunctional monomer.   For example, the local concentration of unreacted monomer for an active center buried within a micro-gel may be considerably different than the mean field concentration. Therefore, the statistical approach is useful for lightly-crosslinked polymer networks formed in reactions containing small concentrations of crosslinking agent (less than a few mole percent) (*49*).   For large quantities of crosslinking agent, highly localized reaction nonidealities such as excessive intramolecular cyclization, heterogeneity, and topological constraints become more important, and the statistical approach breaks down.

Kinetic Descriptions. Kinetic descriptions are based upon the solution of the set of kinetic rate expressions derived from a free radical reaction scheme.   The kinetic approach is closely related to the statistical approach since they are both based upon the mean-field approximation.   Therefore, the two methods will give equivalent results if the analyses are based upon the same set of assumptions.   Kinetic descriptions of free radical crosslinking reactions have been reported by Mikos and collaborators (*67 -69*), and Hamielec and collaborators (*64,70 -72*).

Formulation of the Kinetic Mechanism. We will begin our discussion of the kinetic descriptions by outlining the analysis reported by Mikos *et al* (*67-69*). These authors described the kinetics of free radical crosslinking polymerizations in terms of the concentrations of the mono, di, and pendant vinyl species, and used the method of moments to calculate averages for the linear primary chains. The analysis was based upon a kinetic mechanism which included initiation, propagation and crosslinking, termination by combination and disproportionation, and chain transfer to monomer (*67-69*), but neglected the possibility of intramolecular cyclization (*67-69*). In addition, it was demonstrated that crosslinkers with dependent vinyl group reactivities could be readily included in the analysis (*69*). The most general kinetic reaction scheme (*69*) included seven possible initiation reactions, eighteen possible propagation reactions, and twelve possible termination reactions, as shown below.

The following seven possible initiation reactions were identified (*69*):

$$I \xrightarrow{k_d} 2A \tag{25}$$

$$M_1 + A \xrightarrow{k_{i_1}} P_{1,0,0} \tag{26}$$

$$M_2 + A \xrightarrow{k_{i_2}} Q_{0,1,0} \tag{27}$$

$$P_{p,q,r} + A \xrightarrow{k_{i_3}} P_{p,q-1,r+1} + R_{0,0,1} \tag{28}$$

$$Q_{p,q,r} + A \xrightarrow{k_{i_3}} Q_{p,q-1,r+1}(R_{p,q-1,r+1}) + R_{0,0,1} \tag{29}$$

$$R_{p,q,r} + A \xrightarrow{k_{i_3}} R_{p,q-1,r+1} + R_{0,0,1} \tag{30}$$

$$M_{p,q,r} + A \xrightarrow{k_{i_3}} M_{p,q-1,r+1} + R_{0,0,1} \tag{31}$$

In these equations (*69*), the symbol I refers to the initiator, A corresponds to an initiated radical, while M_1 and M_2 refer to the monovinyl monomer and the divinyl crosslinking agent, respectively. The symbol P, refers to a living polymer chain with a monovinyl terminal group, while Q and R denote living chains terminated by divinyl groups corresponding to a pendant double bond and a crosslink, respectively (*69*). Finally, the symbol M corresponds to a dead polymer chain. Mikos *et al* (*67-69*) used three subscripted indices to describe the composition of the growing polymer chain; these subscripts correspond to the number of monovinyl units (p), pendant vinyl groups (q) and crosslinks (r) in the chain.

With this nomenclature in mind, the meaning of equations 25 through 31 becomes clear. Equation 25 describes the homolytic dissociation of the initiator to produce radicals, while equations 26 and 27 describe the addition of the radicals to the double bonds of the monomer units to produce growing chains of length one. Equations 28 through 31 correspond to addition of the initiator radicals to pendant double bonds on living (equations 28, 29, 30) or dead (equation 31) polymer chains. Note that each of these reactions produce a growing chain containing one crosslink unit (type $R_{0,0,1}$).

A similar analysis yields eighteen distinct possible propagation reactions in which a growing chain propagates with a monomer or a pendant double bond. The first 6 such reactions are depicted below (*69*):

$$P_{p, q, r} + M_1 \xrightarrow{k_{p11}} P_{p+1, q, r} \tag{32}$$

$$P_{p, q, r} + M_2 \xrightarrow{k_{p12}} Q_{p, q+1, r} \tag{33}$$

$$P_{p, q, r} + P_{x, y, z} \xrightarrow{k_{p13}} R_{p, q, r+1} + P_{x, y-1, z+1} \tag{34}$$

$$P_{p, q, r} + Q_{x, y, z} \xrightarrow{k_{p13}} R_{p, q, r+1} + Q_{x, y-1, z+1}(R_{x, y-1, z+1}) \tag{35}$$

$$P_{p, q, r} + R_{x, y, z} \xrightarrow{k_{p13}} R_{p, q, r+1} + R_{x, y-1, z+1} \tag{36}$$

$$P_{p, q, r} + M_{x, y, z} \xrightarrow{k_{p13}} R_{p, q, r+1} + M_{x, y-1, z+1} \tag{37}$$

Note that equations 32 and 33 describe reactions in which a growing chain terminated with a monovinyl unit (type $P_{p,q,r}$) propagates with monomer units, while equations 34 through 37 depict reactions in which the growing chain propagates with pendant double bonds. The other twelve possible propagation reactions are analogous sets of reactions for growing chains of type $Q_{p,q,r}$ and $R_{p,q,r}$.

Termination may occur by combination (six possible reactions), or disproportionation (six possible reactions. The equations that describe termination by combination are shown below (*69*):

$$P_{p, q, r} + P_{x, y, z} \xrightarrow{k_{tc11}} M_{p+x, q+y, r+z} \tag{38}$$

$$P_{p, q, r} + R_{x, y, z} \xrightarrow{k_{tc13}} M_{p+x, q+y, r+z} \tag{39}$$

$$P_{p, q, r} + Q_{x, y, z} \xrightarrow{k_{tc12}} M_{p+x, q+y, r+z} \tag{41}$$

$$Q_{p, q, r} + Q_{x, y, z} \xrightarrow{k_{tc22}} M_{p+x, q+y, r+z} \tag{42}$$

$$Q_{p, q, r} + R_{x, y, z} \xrightarrow{k_{tc23}} M_{p+x, q+y, r+z} \tag{43}$$

$$R_{p, q, r} + R_{x, y, z} \xrightarrow{k_{tc33}} M_{p+x, q+y, r+z} \tag{44}$$

An analogous set of equations describes termination by disproportionation, except that the reaction yields two dead chains equal in length to the original living chains rather than one chain of the combined length (*69*).

Finally, chain transfer during the polymerization is described by 18 reactions (*69*), the first six of which are given below:

$$P_{p, q, r} + M_1 \xrightarrow{k_{f11}} M_{p, q, r} + P_{1, 0, 0} \tag{45}$$

$$P_{p,q,r} + M_2 \xrightarrow{k_{f_{12}}} M_{p,q,r} + Q_{0,1,0} \tag{46}$$

$$P_{p,q,r} + P_{x,y,z} \xrightarrow{k_{f_{13}}} M_{p,q,r} + P_{x,y-1,z+1} + R_{0,0,1} \tag{47}$$

$$P_{p,q,r} + Q_{x,y,z} \xrightarrow{k_{f_{13}}} M_{p,q,r} + Q_{x,y-1,z+1}(R_{x,y-1,z+1}) + R_{0,0,1} \tag{48}$$

$$P_{p,q,r} + R_{x,y,z} \xrightarrow{k_{f_{13}}} M_{p,q,r} + R_{x,y-1,z+1} + R_{0,0,1} \tag{49}$$

$$P_{p,q,r} + M_{x,y,z} \xrightarrow{k_{f_{13}}} M_{p,q,r} + M_{x,y-1,z+1} + R_{0,0,1} \tag{50}$$

Again, the other twelve possible reactions are analogous sets of reactions for growing chains of type $Q_{p,q,r}$ and $R_{p,q,r}$ (*69*).

Derivation of Kinetic Rate Expressions. The above set of mechanistic equations contain thirty distinct rate constants. Based upon this kinetic mechanism, one may readily derive the rate equations for the concentrations of the various polymerizing species (*69*). For example, the rate equation for the total concentration of living polymer is given below (*69*):

$$
\begin{aligned}
r_p = & [k_{i_1}A + (k_{p_{21}} + k_{f_{21}})Q + (k_{p_{31}} + k_{f_{31}})R]M_1 - [2(k_{p_{12}} + k_{f_{12}})M_2 + (k_{p_{13}} + k_{f_{13}})M_3 \\
& + (k_{tc_{11}} + k_{td_{11}})P + (k_{tc_{12}} + k_{td_{12}})Q + (k_{tc_{13}} + k_{td_{13}})R]P
\end{aligned}
\tag{51}
$$

Here the symbols P, Q, and R correspond to the concentrations of total living polymer of each type, as defined below (*69*):

$$P \equiv \sum_{p=0}^{\infty}\sum_{q=0}^{\infty}\sum_{r=0}^{\infty} P_{p,q,r} \tag{52}$$

$$Q \equiv \sum_{p=0}^{\infty}\sum_{q=0}^{\infty}\sum_{r=0}^{\infty} Q_{p,q,r} \tag{53}$$

$$R \equiv \sum_{p=0}^{\infty}\sum_{q=0}^{\infty}\sum_{r=0}^{\infty} R_{p,q,r} \tag{54}$$

Equations similar to equation 51 may be derived for all other reactive species, and, if values of the thirty rate constants can be estimated, the resulting set of coupled differential equations may be solved numerically. Based upon this analysis, the composition of the reaction system may be calculated as a function of time, and structural averages (moments of the distribution) may be obtained for the linear primary polymer chains (*67-69*). This analysis demonstrates the ease with which reaction nonidealities such as unequal reactivities of carbon double bonds may be incorporated into a kinetic analysis of free radical crosslinking polymerizations. However, based upon this kinetic analysis, network chains cannot be distinguished from chains in the sol fraction, and, although the analysis readily provides structural averages for the primary polymer chains (which correspond to the linear chains that would be present if all crosslinks were severed) (67-69), it cannot provide structural information about the larger macromolecular structures including the crosslinks (70).

The Tobita-Hamielec Model and the Crosslink Density Distribution. Kinetic descriptions of free radical crosslinking polymerizations have also been reported by Hamielec and collaborators (70-72). For example, Tobita and Hamielec employed a pseudokinetic rate constant method to simplify the set of kinetic equations to be solved (70-72). The pseudokinetic constants were defined as sums over all possible types of growing polymer radicals and monomers as shown below for the propagation rate constant (71,72):

$$k_p = \sum_{i=1}^{N} \sum_{j=1}^{N} k_{ij} \phi_i^{\bullet} f_i \tag{55}$$

In this equation (71) N denotes the number of monomers in the reaction system, $\Phi_i^{\bullet}$ represents the mole fraction of polymer radical of type i, f_j corresponds to the mole fraction of monomer of type j. Similar equations may be written for chain transfer to monomer (71). Since termination involves the reaction of two radical chains, the expression for the pseudokinetic rate constant takes a slightly different form as illustrated below for termination by combination (71,72):

$$k_{tc} = \sum_{i=1}^{N} \sum_{j=1}^{N} k_{tcij} \phi_i^{\bullet} \phi_j^{\bullet} \tag{56}$$

A similar expression may be written for the rate constant for termination by disproportionation.

The definition of the above pseudokinetic rate constants allows the kinetic analysis of a multicomponent polymerization system to reduce to that of a homopolymerization (71,72). Based upon these definitions, it is clear that the pseudokinetic rate constants are not constant, but change during the course of the polymerization (71). In addition, for the application of the pseudokinetic constants to incur negligible error, the mole fraction of the radicals of each type must not depend appreciably upon chain length (71). Therefore, the following equations must hold (71):

$$\phi_{1,1}^{\bullet} = \phi_{2,1}^{\bullet} = \cdots = \phi_{r,1}^{\bullet} = \cdots = \phi_1^{\bullet} \tag{57}$$

$$\phi_{1,2}^{\bullet} = \phi_{2,2}^{\bullet} = \cdots = \phi_{r,2}^{\bullet} = \cdots = \phi_2^{\bullet} \tag{58}$$

$$\phi_{1,i}^{\bullet} = \phi_{2,i}^{\bullet} = \cdots = \phi_{r,i}^{\bullet} = \cdots = \phi_i^{\bullet} \tag{59}$$

In these equations i again refers to the type of growing radical, and r corresponds to the chain length of the growing radical (71). The term $\Phi^{\bullet}_{r,i}$ corresponds to the mole fraction of radicals of chain length r, while Φ_i has been defined previously. These radical mole fractions are given by the following equations (71):

$$\phi^{\bullet}_{r,i} = \frac{\left[R^{\bullet}_{r,i}\right]}{\sum\limits_{i=1}^{N}\left[R^{\bullet}_{r,i}\right]} = \frac{\left[R^{\bullet}_{r,i}\right]}{\left[R^{\bullet}_{r}\right]} \qquad (60)$$

$$\phi^{\bullet}_{r} = \frac{\sum\limits_{r=1}^{\infty}\left[R^{\bullet}_{r,i}\right]}{\sum\limits_{r=1}^{\infty}\left[R^{\bullet}_{r,i}\right]} = \frac{\left[R^{\bullet}_{i}\right]}{\left[R^{\bullet}\right]} \qquad (61)$$

Tobita and Hamielec have illustrated that the conditions illustrated in equations 57 through 61 typically hold for long polymer chains (71), but may break down for relatively short chains containing fewer than 300 repeating units.

Hamielec and collaborators (64,70-72) illustrated that many reaction nonidealities can be readily included in their kinetic analysis, including unequal reactivity, conversion dependent kinetics, and intramolecular cyclization. An important contribution of the Tobita-Hamielec model (70-72) was the careful accounting for the dependence of the crosslink density on conversion (64). The authors recognized that polymer chains formed at different conversion have different crosslink densities, and accurately accounted for the effect of this crosslink density heterogeneity of the structure of the polymer network (73) as outlined below. The analysis begins with a definition of the crosslink density, ρ, for the primary polymer chains (recall that the primary chains are defined as the linear chains that would be left if all crosslinks were severed).

$$\rho = \frac{(\text{number of crosslinked units})}{(\text{total number of monmeric units bound in the polymer chain})} \qquad (62)$$

Tobita and Hamielec (73) considered a primary polymer chain formed at an arbitrary conversion, b, and divided its crosslink density at a later conversion, n, into two components: the instantaneous crosslink density $\rho_i(b)$ formed by the propagation of the chain with the pendant double bonds of chains formed at conversions earlier than b (64,73), and the additional crosslink density $\rho_a(b,n)$ formed by the propagation of other chains with the pendant double bonds of the chain under consideration (64,73). Note that the additional crosslink density must occur at conversions between b and n. With these definitions, the total crosslink density of the chain under consideration, $\rho(b,n)$, is given by the following equation (73).

$$\rho(b, n) = \rho_i(b) + \rho_a(b, n) \qquad (63)$$

Tobita and Hamielec derived the following equation for the additional crosslink density (*73*):

$$\frac{\partial \rho_a(b,n)}{\partial n} = \frac{k_p^{*0}(n)\left[F_2(b) - \rho_a(b,n) - \rho_c(b,n)\right]}{k_p(n)[M]_n V(n)} \tag{64}$$

In this equation (*73*), $F_2(b)$ denotes the mole fraction of divinyl monomer bound in the primary polymer chains, $[M]_n$ is the monomer concentration at conversion n, $V(n)$ is the reaction volume at this conversion, and pseudokinetic constants are given by the following equations (*73*):

$$k_p^{*0} = k_{p13}^* \phi_1^\bullet + k_{p23}^* \phi_2^\bullet + k_{p33}^* \phi_3^\bullet \tag{65}$$

$$k_p = \left(k_{11}f_1 + k_{12}f_2\right)\phi_1^\bullet + \left(k_{21}f_1 + k_{22}f_2\right)\phi_2^\bullet + \left(k_{31}f_1 + k_{32}f_2\right)\phi_3^\bullet \tag{66}$$

Finally, because all additional crosslinks must have an instantaneous partner, the instantaneous crosslink density at conversion u may be given by the following equation (*73*).

$$\overline{\rho}_i(u) = \int_0^u \frac{\partial \rho_a(b,u)}{\partial u}\, db \tag{67}$$

Therefore, using this set of equations it is possible to calculate the crosslink density as a function of the birth conversion of the primary polymer chains. Results of this analysis are shown in Figure 2 (*71*). The figure illustrates the crosslink density distribution for a system in which the pendant double bonds have reduced reactivity and there is no cyclization (*71*). The dashed bottom line corresponds to the instantaneous crosslink density $\rho_i(b)$, while the distance between curves for different conversions (b and n) corresponds to the additional crosslink density $\rho_a(b,n)$. The figure illustrates that although the crosslink density is rather uniform at low conversions (for example n=0.3), it becomes a strong function of the birth conversion at high values of conversion (*71*). This outline should provide the reader with an appreciation of the ingenious analysis developed by Hamielec and collaborators to account for the conversion dependence of the crosslink density. For more details, the reader is referred to reference, *71* and *73*.

Advantages and Limitations of the Kinetic Approach. Kinetic descriptions have some distinct advantages for modeling of free radical crosslinking polymerizations containing relatively small concentrations of crosslinking agents. The greatest advantage arises form the fact that these models may accurately account for the history-dependent effects of these kinetically-controlled reactions (*64, 70-72*). These features are more readily incorporated into a kinetic description of the reaction than the statistical descriptions described earlier. However, a disadvantage of kinetic models relative to the statistical descriptions is the difficulty of obtaining structural information about the network (*59,60,65*). Although we have seen that structural

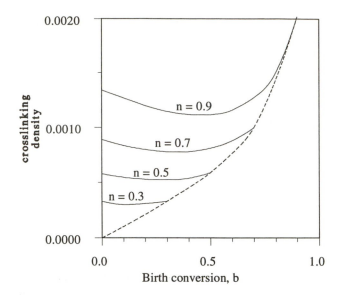

Figure 2. Representative crosslink density distribution from the Tobita-Hameliec analysis. (Adapted from ref. 71.)

information such as the crosslink density distribution may be obtained using kinetic models (*64*), the technique has no analog to the statistical extinction probability, and therefore cannot provide the same information about the post-gel structure as the statistical descriptions (*60,64,65*). Again, a combined kinetic/statistical approach may provide the most information while accurately accounting for the history-dependent effects (*64*). For example, Zhu and Hamielec (*64*) reported a clever analysis in which the Tobita-Hamielec crosslink density distribution was combined with the statistical description of the free radical crosslinking reactions. Like the statistical approach, the kinetic method is based upon the mean field approximation, and therefore cannot accurately account for topological constraints and the corresponding spatial heterogeneities which lead to highly localized effects (see the kinetic gelation discussion below). Finally, kinetic descriptions may be difficult to implement if reliable values of the kinetic constants are not available.

Kinetic Gelation Simulations. The third type of model for free radical crosslinking polymerizations are the kinetic gelation simulations. These simulations are based upon computer-generated random walks, typically on a cubic lattice. The discussion below will illustrate that kinetic gelation simulations currently have limited applicability to lightly crosslinked systems, therefore our consideration of these models will be somewhat abbreviated. However, because these relatively new models have already established a niche in the description of highly crosslinked systems, and will only continue to be refined and improved, it is important to provide an overview of the kinetic gelation approach.

The first kinetic gelation model was reported by Manneville and de Seze, (*74*) who distributed bi-functional and tetra-functional monomer units randomly on a cubic lattice, then started the simulation by allowing a fraction of the sites to become active centers and propagate through the lattice. At each step during the simulation an active center was chosen at random and allowed to chose one of its first or second nearest neighbors to react with (75). If the chosen site was an unreacted functionality, a bond was formed and the active center was transferred; if the chosen site possessed another active center, the radicals annihilated each other and termination was achieved. Finally, if an active center became surrounded only by reacted sites, the radical had no units to react with and became entrapped. Structural information was obtained by stopping the simulation at various conversions and examining the ensemble of polymer chains that had been formed. Manneville and de Seze reported that their model incorporated two drastic simplifications (74): constraining the system to a cubic lattice, and ignoring the possibility of molecular motion.

The kinetic gelation approach has been refined and modified by several authors. Herrmann and collaborators (76 -79) included the effects of inert solvents and monomer mobility in an approximate manner. Boots and collaborators (80 -82) relaxed the assumptions of instantaneous and simultaneous initiation by allowing initiation to occur at a much slower rate than termination. Simon et al. (83) allowed an exponentially decaying rate of initiation. Finally, Bowman and Peppas (84) included most of these effects as well as improved descriptions of monomer size and mobility. Most of the kinetic gelation simulations reported to date have been based upon simple cubic lattices, and have employed periodic boundary conditions to reduce the required lattice size.

A representative kinetic gelation simulation is shown in Figure 3 (81). The figure contains a two-dimensional 30 x 30 square lattice that contains 100% tetrafunctional monomers at a double bond conversion of 0.58. Figure 3 illustrates one of the key strengths of the kinetic gelation approach: the ability to account for reaction hetero- geneity. Because the reaction occurs primarily in the vicinity of the active centers, some areas of the lattice show no reaction while others are heavily crosslinked. Therefore, the kinetic gelation approach is not based upon the mean field approximation, and may account for reaction heterogeneity, microgel formation, and localized effects.

The ability of kinetic gelation simulations to account for localized effects is illustrated in Figure 4. These results, taken from a paper by Bowman and Peppas (84), were obtained from a simulation of a tetrafunctional homopolymerization on a three-dimensional 20 x 20 x 20 lattice. Here the reactivity ratio is defined as the reactivity of the pendant vinyl groups divided by that of the monomer functional groups. The figure illustrates that at very low conversions, the reactivity of the pendant double bonds is much higher than that of the monomer units. This effect arises from the fact that for very low conversions, the local radical concentration is high in the vicinity of the pendant double bonds (much higher than the average over the whole reaction system). This locally high concentration dramatically increases the probability that the pendant double bond will react relative to the average monomer unit. As the conversion is increased, the reactivity decreases, and eventually falls below unity as the pendant double bonds become inaccessible to the active centers.

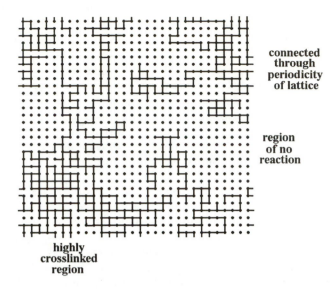

Figure 3. Representative two dimensional kinetic gelation simulation illustrating heterogeneity. (Reproduced with permission from ref. 82. Copyright 1985 Elsevier Science Publishing).

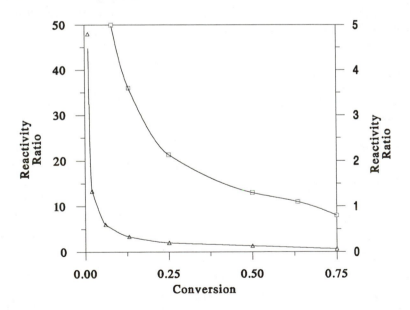

Figure 4. Reactivity ratio versus conversion of a homopolymerization of a tetrafunctional monomer: Δ = reactivity ratio from 0 to 50, □ = reactivity ratio expanded from 0 to 5. (Reproduced with permission from ref. 84. Copyright 1992 Pergamon Press, Inc.)

Advantages and Limitations of Kinetic Gelation Simulations.   Although kinetic gelation models can account for local effects and reaction heterogeneity, they are presently unable to account for many of the molecular phenomena which can significantly affect the polymerization characteristics.   As recognized by Manneville and de Seze in their original paper, (*74*) most of the limitations of the models arise from the choice of a lattice, and the inability to properly account for molecular motions.   For example, constraining the system to a lattice limits the number of units an active center may react with, and significantly distorts the bond angles in the polymer backbone chain (this will affect probabilities of cyclization, etc.).   The inability to account for monomer and polymer mobility precludes the possibility of accurately including diffusion-controlled reactions.   Boots and Dotson (*85*) illustrated that failure to account for polymer mobility results in an overestimation of the heterogeneity in the reaction.   Therefore the kinetic gelation approach is most useful for describing highly crosslinked systems formed with large amounts of tetrafunctional crosslinking agent, and is of limited value for describing lightly crosslinked superabsorbing polymers.

The inability of current kinetic gelation simulations to accurately account for molecular motions in lightly crosslinked systems is illustrated by the unreasonably high number of trapped active centers predicted by the simulations.   For example, in a simulation for a system containing 90% monovinyl and 10% divinyl monomer (*77*), all of the active centers became trapped for a limiting conversion of less than 60%.   Relatively low limiting conversions and high fractions of trapped active centers have been experimentally observed for polymerizations containing high concentrations of tetrafunctional crosslinking agent, and again the kinetic gelation simulations do a better job describing these systems.   However, in contrast to the experimental trends, the kinetic gelation simulations predict that radical trapping will become more pronounced as the concentration of the tetrafunctional monomer is decreased, clearly demonstrating the deficiencies of the method for describing lightly crosslinked systems which contain only a few mole percent of crosslinking agent.

Conclusions

Theoretical descriptions of free radical crosslinking polymerizations may be classified as statistical models, kinetic descriptions, or kinetic gelation simulations.   The above discussion has illustrated that each approach offers a distinct set of strengths and limit-ations.   The greatest strength of the statistical approach arises from the wealth of post-gel structural information that may be obtained.   For example, through the use of the statistical extinction probability the gel fraction may be readily distinguished from the sol fraction, elastically active chains may be distinguished from dangling chain ends, etc.   However, it is difficult to account for history-dependent effects using the statistical approach, and the method breaks down for highly crosslinked systems due to localized effects and topological constraints.   The greatest advantage of the kinetic descriptions is the ease with which history-dependent reaction non-idealities may be described, as illustrated by the crosslink density distribution analysis of Tobita and Hamielec.   A disadvantage of kinetic descriptions is the difficulty of obtaining structural information about the polymer network (*59,60,65*).   Therefore, the most useful modeling approach may be a combination of the kinetic and statistical methods in which the statistical weights are derived by solution of the kinetic equations, or the kinetic crosslink density distribution is combined with a statistical description. Both the statistical and kinetic approaches are based upon the mean-field approximation and are unable to describe the topological constraints and local effects prevalent for systems containing high concentrations of tetrafunctional monomer. The kinetic gelation simulations avoid the mean field approach and may account for heterogeneity and localized effects. However the method is of limited value for lightly crosslinked polymers due to its inability to accurately account for molecular motions. This approach is most useful for highly crosslinked systems containing a preponderance of tetrafunctional monomer.

Literature Cited

1. Gross, J.R. In *Absorbent Polymer Technology*; Brannon-Peppas, L.; Harland, R.S. Eds.;. Elsevier Science Publishing Company Inc.: New York, NY, 1990; pp. 3-22.
2. Buchholz, F.L. In *Absorbent Polymer Technology*; Brannon-Peppas, L.; Harland, R.S. Eds.; Elsevier Science Publishing Company Inc.: New York, NY, 1990; pp. 23-44.
3. Samsonov, G.V.; Kuznetsova, N.P. *Advances in Polym. Sci.* **1992,** *104*, 1.
4. Kazanskii, K.S.; Dubrovskii, S.A. *Advances in Polym. Sci.* **1992,** 104, 97.
5. Colombo P., *Adv. Drug Delivery Rev.* **1993,** *11*, 37.
6. Brannon-Peppas, L. In *Absorbent Polymer Technology*; Brannon-Peppas, L.; Harland, R.S. Eds.; Elsevier Science Publishing Company Inc.: New York, NY, 1990; pp. 45-66.
7. Buchholz, F.L. In *Absorbent Polymer Technology*; Brannon-Peppas, L.; Harland, R.S. Eds.; Elsevier Science Publishing Company Inc.: New York, NY, 1990; pp. 233-247.
8. Yin, Y.L.; Prud'homme, R.K.; Stanley, F. In *Polyelectrolyte Gels*; Harland, R.S.; Prud'homme, R.K. Eds., ACS Symposium series 480, ACS: Washington, D.C., 1992; pp. 91-113.
9. Peppas, N.A.; Mikos A.G. In *Hydrogels in Medicine and Pharmacy Vol. I*; Peppas, N.A. Ed.; CRC Press: Boca Raton, Florida, 1986; pp 1-*25*.
10. Schosseler F.; Ilmain, F.; Candau, S.J. *Macromolecules*, **1991**, *24*, 225.
11. Peppas, N.A. *Hydrogels in Medicine and Pharmacy*; CRC Press: Boca Raton, FL, 1987.
12. Langer, R.S.; Peppas, N.A. *Biomaterials* **1981**, *2*, 201.
13. Peppas, N.A. In *Biomaterials: Interfacial Phenomena and Applications*; Cooper, S.L.; Peppas, N.A., Eds.; Advances in Chem. Series; ACS: Washington, D.C., 1982; vol. 199.
14. Buchholz, F.L. *Poly. Mater. Sci. Eng.* **1993**, *69; 489-490*.
15. Hunkeler, D.; Hameilec, A.E. In *Polyelectrolyte Gels: Properties, Preparation, and Applications*; Harland, R.S.; Prud'homme, R.K., Eds.; ACS Symposium Series; ACS: Washington, D.C., 1992; Vol. 480, pp. 53-89.
16. Chapiro, A; Dulieu, J. *Eur. Polym. J.* **1977**, *13*, 563.
17. Bajoras, G.; Makuska, R. *Polym. J.* **1986**, *18*, 955.
18. Kabanov, V.A.; Topchiev, D.A.; Karaputadze, M. *J. Polym. Sci.* **1973**, *42*, 173.

19. Kabanov, V.A.; Topchiev, D.A.; Karaputadze, M.; Mkrtchian, L.A. *Eur. Polym. J.* **1975**, *11*, 153.
20. Chapiro, A. *Pure Appl. Chem.* **1972**, *30*, pp. 77.
21. M. Gordon; Roe, R. J. *J. Polym. Sci.*, **1956**, *21*, 27.
22. Story, B. T. *J. Polym. Sci. A*, **1965**, *3*, 265.
23. Malinski, J.; Klaban, J.; Dusek, K. *J. Macromol. Sci. Chem.* **1971**, *A5*, 1071.
24. Dusek, K *Polym. Bull.*, **1980**, *3*, 19.
25. Dusek, K; Spevacek, J. *Polymer*, **1980**, *21*, 750.
26. Whitney, R. S.; Burchard, W. S. *Macromol. Chem.*, **1980**, *181*, 869.
27. Landin, D. T.; Macosko C. W. *Macromolecules*, **1988**, *B-21*, 846.
28. Matsumoto, A.; Yonezawa, S.; Oiwa, M. *Eur. Polym. J.*, **1988**, *24*, 703.
29. Matsumoto, A.; Matsuo, H.; Ando, H.; Oiwa, M. *Eur. Polym. J.*, **1989**, *25*, 237.
30. Shah, A.C.; Holdaway, I.; Parsons, I.W.; Hayward, R.N., *Polymer*, **1978**, *19*, 1067.
31. Dusek, K.; Spevacek, J. *J. Polym. Sci. Polym. Phys. Ed.* , **1980**, *18*, 2027.
32. Shah, A.C.; Parsons, I.W.; Hayward, R.N. *Polymer* **1980**, *21*, 825.
33. Leicht, R.; Fuhrmann, J. *Polym. Bull.*, **1981**, *4*, 141.
34. Dusek, K *Developments in Polymerization*, Applied Science Publishers, London, Vol. 3, 1982.
35. Galina, H.; Dusek, K.; Tuzar, Z.; Bohanecky, M.; Stokr, J. *Eur. Polym. J.*, **1980**, *16*, 1043.
36. Dusek, K.; Spevacek, J. *J. Polym. Sci., Polym. Phys.*, **1980**, *3*, 473.
37. Pilar, J.; Horak, D.; Labsky, J.; Svec, F. *Polymer*, **1988**, *29*, 500.
38. Tanaka, H.; Fukumori, K.; Kakurai, T *J. Chem. Phys.*, **1988**, *89*, 3363.
39. Odian, G. *Principles of Polymerization*; John Wiley and Sons: New York, NY 1991.
40. Scranton, A.B.; Bowman, C.N.; Klier, J.; Peppas, N. *Polymer*, **1992**, 33, 1683-1689.
41. Ulbrich, K.; Ilavsky, M.; Dusek, K.; Kopecek, J. *Eur. Polym. J.* **1977**, *13*, p. 579.
42. Ulbrich, K.; Dusek, K.; Ilavsky, M.; Kopecek, J. *Eur. Polym. J.* 1978, *14*, p. 45.
43. Kamachi, M. *Adv. Polym. Sci.* **1987**, *82*, 207.
44. Shen, J.; Tian, Y. *Macromol. Chem., Rapid Commun.*, **1987**, *8*, 615.
45. Zhu, S.; Tian, Y.; Hamielec, A.E. *Polymer*, **1990**, *31*, 154.
46. Dotson, N.A.; Galvan, R.; Macosko, C.W. *Macromolecules*, **1988**, *21*, 2560.
47. Landin, D.T.; Macosko, C.W. In *Characterization of Highly Crosslinked Polymers*; Labana, S.S. Ed.; ACS Symposium Series; ACS: Washington, D.C., 1984, Vol. 243; pp. 22-46.
48. Williams, R.J.J.; Vallo, C.I. *Macromolecules*, **1988**, *21*, 2571.
49. Scranton, A.B.; Peppas, N.A., *J. Polym. Sci.*, **1990**, *28*, pp. 39-57.
50. Scranton, A.B.; Klier, J.; Peppas, N.A. *Macromolecules*, **1991**, *24*, p. 1412.
51. Dotson, N.A. *Macromolecules*, **1992**, *25*, 308.
52. Gordon, M. *Proc. R. Soc. London Ser. A*, **1962**, *268*, p. 240.
53. Gordon, M.; Malcolm, G.N. *Proc. R. Soc. London Ser. A*, **1966**, *295*, 29.
54. Dobson, G.R.; Gordon, M. *J. Chem. Phys.*, **1965**, *43*, 705.
55. Dusek, K *Adv. Polym. Sci.*, **1986**, *78*, 1.
56. Gordon, M. *Proc. R. Soc. London Ser. A*, **1962**, *268*, 240.

57. Gordon, M.; Malcolm, G.N. *Proc. R. Soc. London Ser. A*, **1966**, *295*, 29.

58. Dobson, G.R.; Gordon, M. *J. Chem. Phys.*, **1965**, *43*, 705.

59. Dusek, K. *Brit. Polym. J.*, **1985**, *17*, 185.

60. Dotson, N.A.; Macosko, C.W.; Tirrell, M. *Proceed. Top. Confer. Emerg. Techn. Mater., AICHE*, **1989**, *2*, 659.

61. Dusek, K. *Macromolecules*, **1984**, *17*, 716.

62. Miller, D.R.; Macosko, C.W. In *Biological and Synthetic Polymer Networks*, O. Kramer, Ed.; Elsevier Applied Science: New York, NY, 1988, pp. 219-231.

63. Tsou, A.H.; Peppas, N.A. *J. Polym. Sci.*, **1988**, *26*, 2043.

64. Zhu, S.; Hamielec, A.E. *Macromolecules* **1992**, *25*, 5457.

65. Dotson, N.A. *Macromolecules*, **1989**, *22*, 3690.

66. Jacobs, J.L.; Scranton, A.B. *Polymer News*, **1994**, *19*, 80.

67. Mikos, A.G.; Takoudis, C.G.; Peppas, N.A. *Macromolecules*, **1986**, *19*, 2174.

68. Mikos, A.G.; Peppas, N.A. *J. Controlled Release*, **1986**, *5*, 53.

69. Mikos, A.G.; Takoudis, C.G.; Peppas, N.A. *Polymer*, **1987**, *28*, 998.

70. Tobita, H.; Hamielec, A.E. *Polymer*, **1991**, *32*, 2641.

71. Tobita, H.; Hamielec, A.E. *Macromolecules*, **1989**, *22*, 3098.

72. Tobita, H.; Hamielec, A.E. *Makromol. Chem., Makromol. Symp.*, **1988**, *20/21*, 501.

73. Tobita, H.; Hameilec, A.E. *Polymer*, **1992**, *33*, 3647.

74. Manneville, P.; de Seze, L. In *Numerical Methods in the Study of Critical Phenomena*; della Dora, I.; Demongeot, J.; Lacolle, B.; Eds. Springer, Berlin, 1981, pp. 116-124.

75. Manneville, P.; de Seze, L. In *Numerical Methods in the Study of Critical Phenomena*; della Dora, I.; Demongeot, J.; Lacolle, B.; Eds. Springer, Berlin, 1981, pp. 116-124.

76. Herrmann, H.J.; Landau, D.P.; Stauffer, D. *Phys. Rev. Lett.*, **1982**, *49*, 412.

77. Herrmann, H.J.; Stauffer, D.; Landau, D.P. *J. Phys. Math. Gen.*, **1983**, *16*, 1221.

78. Matthews-Morgan, D.; Landau, D.P.; Herrmann, H.J. *Phys. Rev.*, **1984**, *B11*, 6328.

79. Bansil, R.; Herrmann, H.J.; Stauffer, D. *Macromolecules*, **1984**, *17*, 998.

80. Kloosterboer, J.G.; van de Hei, G.M.M.; Boots, H.M.J. *Polym. Commun.*, **1984**, *25*, 354.

81. Boots, H.M.J.; Pandey, R.B. *Polym. Bull.*, **1984**, *11*, 415.

82. Boots, H.M.J.; Kloosterboer, J.G.; van de Hei, G.M.M. *Brit. Polym. J.*, **1985**, *17*, 219.

83. Simon, G.P.; Allen, P.E.M.; Bennett, D.J.; Williams, D.R.G.; Williams, E.H. *Macromolecules*, **1989**, *22*, 3555.

84. Bowman, C.N.; Peppas, N.A. *Chem. Eng. Sci.*, **1992**, *47*, 1411.

85. Boots, H.M.J.; Dotson, N.A. *Polym. Comm.*, **1988**, *29*, 346.

RECEIVED June 16, 1994

Chapter 2

Preparation Methods
of Superabsorbent Polyacrylates

Fredric L. Buchholz

**Specialty Chemicals Research and Development Department,
1603 Building, Dow Chemical Company, Midland, MI 48674**

Superabsorbent polymers, used in infant diapers to absorb body fluids, are prepared from acrylic acid and a crosslinker, in aqueous solution or in a suspension of the aqueous solution in a hydrocarbon. The product is partially neutralized either before or after the polymerization step. Combinations of redox and thermal free-radical initiators, chelating agents, chain-transfer agents, and grafting agents are used to control the polymerization kinetics and molecular characteristics of the network. The key properties of swelling capacity and gel modulus are controlled by the choice and amount of cross-linker. The gel-like polymer mass resulting from the polymerization is dried in continuous ovens, and the dry polymer milled to the desired size. Particle size of the product from suspension polymerization is controlled by choice of suspending agent.

The commercially important superabsorbent polymers are cross-linked polymers of partially neutralized acrylic acid. They are formally terpolymers of acrylic acid, sodium acrylate, and a cross-linker. Some of the commercially available polymers are graft terpolymers with starch or poly(vinyl alcohol) as the graft substrate. The swelling and elasticity of these polymers depend on the precise structure of the polymer network and primarily on the cross-link density. The techniques for preparing polyacrylate superabsorbents that are described in the literature are all aimed at adjusting the balance of properties of the superabsorbent, through control of the network structure.

The focus of this chapter is the preparation and manufacture of the commercially available materials. Several important techniques used to alter the properties will be described along with some important aspects of the chemical processes employed. First, however, some of the techniques used to measure the properties of superabsorbent polymers will be described in brief.

0097–6156/94/0573–0027$08.00/0
© 1994 American Chemical Society

Characterization of Superabsorbent Polyacrylates

The key properties of superabsorbent polymers are the swelling capacity and the elastic modulus of the swollen, cross-linked gel. These properties of the product both are related to the cross-link density of the network: modulus increases and swelling capacity decreases with increasing cross-link density. The cross-link density of superabsorbent polymers used in personal care applications is very low, about 0.03 mole of cross-linker per liter of dry polymer. Because of this very low concentration, conventional spectroscopic techniques, such as infrared spectroscopy and nuclear magnetic resonance spectrometry, are not very useful for characterizing the cross-link density. Mostly, gravimetric methods are used to measure the swelling capacity, and rheological methods are used to measure the modulus. The cross-link density can be estimated from these measurements by applying an appropriate mathematical model of the relationship between network structure and properties.

The most common measure of the swelling capacity that is used in the superabsorbent polymer industry is the centrifuged capacity (1). A porous bag, measuring about 3 inches square, is constructed from a heat-sealable, water-wettable fabric. A small quantity of the granular polymer powder is put into the bag, and the bag is sealed. The bag is then immersed in a bath of the desired test fluid and left to absorb liquid for a given length of time. If the kinetics of swelling are to be measured, bags may be withdrawn from the bath at different times. If equilibrium swelling capacity is to be approximated, the bags are left in the bath for 30-90 minutes, depending on the particle size of the individual particles and their swelling rate, and then withdrawn. The bags are placed into a laboratory centrifuge, equipped with a perforated spin-basket, and centrifuged for a few minutes to remove any unabsorbed fluid from between the particles of the gel mass. Alternatively, the excess liquid may be removed by blotting with a porous paper. The swelling capacity is calculated from the increase in mass of the polymer sample and is typically reported as a ratio of the grams of fluid absorbed per gram of dry polymer. The volume fraction of polymer in the swollen gel, used as a measure of the extent of swelling in most university literature, is easily calculated knowing densities of both the polymer and fluid.

The elastic modulus of swollen gels is typically measured on a compact mass of swollen particles, by means of an oscillatory stress rheometer (2). The swollen mass of particles is packed into the space between the plates of the rheometer, and an oscillating shear stress is applied to one plate. The motion of the opposite plate of the rheometer is measured and is related to the damping of the mechanical wave passing through the sample. The shear modulus is calculated from the measurements. Similar measurements have been made on single disks of gel cut from a larger sample or molded into the required cylindrical shape.

A few other tests have been developed to measure fluid absorption under conditions that simulate those in the actual personal-use application. For example, the polymer in a diaper will occasionally be under a compressive load when a baby sits or lies on the diaper. The absorbency under load test (3) measures the swelling capacity of the polymer while an external pressure is applied to the swelling gel. Similar information has been collected on water-swellable gels by measurements of the swelling pressure, which is the pressure generated by the gel in contact with an external source of fluid while being confined to a given volume (*4*). This information is related to the osmotic compressibility of the network, which is a quantity relating the amount of change in the swelling pressure to the change of polymer concentration in the gel. Stated another way, the osmotic compressibility is a measure of how easily a given gel can be deswollen by an external pressure (*5*).

A few methods have been developed to measure the results of the extent of reaction and completeness of cross-linking. The amount of unreacted monomer in a sample may serve as a measure of the extent of the polymerization reaction. In this technique, the monomer is extracted from the network and quantitated by means of liquid chromatography (*6*). The completeness of cross-linking can be estimated by measuring the amount of soluble polymer formed during the polymerization. This is done by extracting the soluble polymer from the network and quantitating it by titration of the carboxylic acid groups (*2*), or by gravimetry after drying the extract (*1*). The degree of neutralization of the carboxylic acid groups can be determined using a technique employing a metal-ion specific electrode (*7*).

Free-radical Polymerization in Aqueous Solution

Superabsorbents are prepared by free-radical initiated polymerization of acrylic acid and its salts, with a cross-linker, in aqueous solution or in suspensions of aqueous solution drops in a hydrocarbon. These two principal processes, bulk solution polymerization and suspension polymerization, share many features. The monomer and cross-linker concentrations, the initiator type and concentrations, polymerization modifiers, the relative reactivities of the monomers, the basic polymerization kinetics and the reaction temperature are all significant factors in both processes.

Monomer Concentration. The concentration of monomer in the reaction solution affects the properties of the resulting polymer, the kinetics of the reaction and the economics of the process. High monomer concentration results in increasing toughness of the intermediate gel polymer as the polymerization progresses. The toughness of the gel affects the design of the equipment, the size of gel particles produced during agitation of the reaction mass and the method of heat removal. In addition, chain transfer to polymer increases with monomer concentration, especially at high extent of conversion, and this results in increasing amounts of branching and self-crosslinking reactions that affect product properties. Chain-transfer agents are useful to combat these side-reactions (*8*). Another factor influencing the choice of monomer concentration is that the efficient use of the cross-linker increases with monomer concentration because the solubility of cross-linkers—typically not very water soluble—often increases with monomer concentration due to the increasing organic content of the monomer phase. Also, network cyclization reactions decrease at higher monomer concentration (*9*).

A factor of considerable importance to the polymerization in large quantity and at high monomer concentration is the large heat of polymerization of acrylic acid. The monomer yields 18.5 kcal/mole upon polymerization, making temperature control important. Lower monomer concentration lessens the potential adiabatic temperature rise but also lowers the volumetric efficiency of the reaction equipment and affects the polymerization kinetics. Evaporative cooling at reduced pressure may be used to remove the heat of polymerization (*10*). The heat has also been used to dry the polymer to a foamy mass (*11*). If the suspension process is used, traditional methods of heat transfer can be used because a lower viscosity, liquid state is maintained. This can be an advantage when precise temperature control is desired.

Because acrylic acid is stored inhibited against premature polymerization with p-methoxyphenol and oxygen, either the p-methoxyphenol or the oxygen must be removed from the monomer solution before polymerization will proceed. The simplest method is to strip the dissolved oxygen from the solution with a stream of nitrogen gas. Alternatively, the oxygen may be reacted from the solution by a metal-ion catalyzed reaction that forms hydroperoxides from the oxygen and monomer.

Initiators. The polymerization is initiated by free-radicals in the aqueous phase, using thermally decomposable initiators, redox initiators or combinations. Redox systems used for the cross-linking copolymerizations include couples of persulfate/bisulfite, persulfate/thiosulfate, persulfate/ascorbate and hydrogen peroxide/ascorbate. Thermal initiators include persulfates, 2,2'-azobis-(2-amidinopropane)-dihydrochloride, and 2,2'-azobis(4-cyanopentanoic acid). Combinations of initiators are used when the polymerization takes place over a broad temperature range. In this case, it may be desirable to maintain a constant rate despite the change in temperature. Appropriate concentrations of multiple initiators can achieve the desired constant-rate polymerization (*12*).

In graft copolymerization of vinyl monomers to polysaccharide substrates, initiation is accomplished using a redox reaction of an oxidant, such as the oxidized form of a metal ion, with oxidizable groups of the polysaccharide. When the metal ion is reacted with the graft substrate before monomer is added, grafting efficiency is increased (*13*). In addition to initiating the polymerization reaction, initiators are a factor in reducing the levels of unreacted monomer during the drying step (*14*) and can contribute to undesirable chain cleavage reactions that occur when the gel is handled at higher temperatures. For example, higher content of soluble polymer is found when sodium polyacrylate gels made with ammonium persulfate initiator are dried in a very hot oven (*15*).

Neutralization. The monomers and cross-linker are dissolved in water at a desired concentration, usually from about 10% - 70%. The acrylic acid usually is partially neutralized before the polymerization is initiated (*16*), but the cross-linked polymer can be neutralized after polymerization is complete (*2, 17*). In suspension processes, neutralization of the monomer is required due to the appreciable solubility of acrylic acid in hydrocarbons (continuous phase).

Inexpensive bases, such as sodium hydroxide and sodium carbonate, are used as neutralizing agents. A choice would be made based on consideration of the pH of the base solution and the resulting potential for hydrolyzing cross-linker, the solubility limits of the base in water and on the solubility of the monomer salt in water. For example, potassium acrylate is more soluble in water than is sodium acrylate.

Other Polymerization Additives. Chain transfer agents may be used to control network properties through control of polymer backbone molecular mass. A variety of chain transfer agents are known for water soluble monomer systems (*18*). Examples are mercapto- compounds (*19-22*), formic acid (*23*), carbon tetrachloride, isopropanol (*24*), monobasic sodium phosphate and hypophosphite salts (*25*). Some of these also have been used in cross-linked systems (*26-28*). Typically, a higher swelling capacity is obtained when chain transfer agents are used, as predicted by theory (*9*). Chain-transfer agents and other radical scavengers can also help prevent oxidative degradation of the polymer after it has been hydrated during use (*29*). In addition, chain-transfer agents may be used to minimize the branching and self-crosslinking reactions that have been reported during polymerization at higher monomer concentration (*30*).

Chelating agents are useful to help control the variable concentrations of metal ions that are present in the water used as reaction solvent. These metal ions, notably iron, catalyze many free-radical reactions. When their concentration is variable, initiation is irregular and possibly uncontrollable. Metal ions can also catalyze reactions of the initiators that lead to non-radical products; this wastes the initiator and can cause incomplete conversion of the monomer to polymer.

Kinetics. The polymerization kinetics are affected by monomer and initiator concentration, by pH (*31-33*) and by the ionic strength of the reaction medium (*34*). Most, if not all, of the commercial processes use persulfate salts as one of the initiators. In this case, the kinetics of the polymerization are proportional to the $3/2$ power of acrylic acid concentration and to the square root of the concentration of persulfate. Polymerization rates decrease with increasing extent of neutralization, but this effect is moderated in industrial processes by the high ionic strength of the monomer solution when the acrylic acid is partially neutralized and at high concentration.

Cross-linkers. Relatively small amounts of cross-linkers play the major role in modifying the properties of superabsorbent polymers. The cross-linkers typically used in superabsorbent polymers are di- and tri-acrylate esters such as 1,1,1-trimethylolpropanetriacrylate or ethylene glycol diacrylate. In addition to modifying the swelling and mechanical properties, the cross-linker affects the amount of soluble polymer formed during the polymerization. The tendency of a cross-linker to be depleted earlier in the polymerization is reflected in its reactivity ratio with acrylic acid or sodium acrylate. Early depletion of cross-linkers should cause higher soluble fraction in the product.

Unfortunately, reactivity ratios have not been directly determined for the cross-linkers typically used in making superabsorbent polymers. However, these can be estimated from the reactivity ratios of structurally analogous monomers, using the Alfrey-Price Q-e scheme. This has been done for several cross-linker analogs, using the available Q-e values (*35*). The results, shown in Table I, suggest that cross-linkers similar to triallylcitrate should yield polymers with a lower amount of soluble polymer while cross-linkers similar to ethyl methacrylate should yield higher amounts of soluble polymer. This prediction is confirmed in the results from polymerizations, shown in Table II.

Table I. Reactivity Ratios for Structural Analogs of Cross-linkers

Monomer /Cross-linker Analog	r_1	r_2
1. Acrylic acid		
triallylcitrate	5.636	0.049
acrylamide (bisacrylamides)	2.676	0.324
ethyl acrylate (diacrylates)	1.514	0.576
ethyl methacrylate (dimethacrylates)	0.585	1.024
2. Sodium acrylate		
acrylamide	2.852	0.355
ethyl acrylate	1.598	0.631
ethyl methacrylate	0.902	1.173

Table II.   Cross-linker Reactivity Effect on Gel Fraction

Cross-linker	$\sim r_1{}^a$	Gel fraction[b]
Methylenebisacrylamide	2.8	0.983
Triallylcitrate	2.8	0.981
Ethyleneglycol diacrylate	1.56	0.955
Ethyleneglycol dimethacrylate	0.74	0.789

[a]Taken as the average of the r_1 values for acrylic acid and sodium acrylate
with the respective cross-linker analogs
[b]Identical polymerizations of 65 mole% neutralized acrylic acid with 0.145 mole%
cross-linker, in aqueous solution at 32 mass% monomer, initiated with a sodium
persulfate and sodium erythorbate redox couple at 28 °C.

    The choice of cross-linker will also depend on the method used to neutralize the carboxylic acid groups. A high pH process step may require a hydrolytically stable cross-linker, such as tetraallyloxyethane, rather than a diacrylate ester. In a suspension polymerization process, the availability of the cross-linker in the aqueous phase will be controlled by the partition coefficient of the cross-linker between the aqueous phase and the hydrocarbon, continuous phase. The partition coefficient will depend on the extent of neutralization and on the nature of the hydrocarbon (eg., whether aromatic or aliphatic). The solubility of the cross-linker in the monomer solution also affects the efficiency of cross-linking in solution polymerization (36). Efficiency of cross-linking will also depend on steric hindrance and reduced mobility at the site of a pendent double bond, the tendency of a given cross-linker to undergo intermolecular addition (cyclopolymerization) (*37*) and the solubility of the cross-linker that can depend on the extent of neutralization, as noted above. As a result of a combination of factors such as these, different cross-linkers can exhibit much different effectiveness in the cross-linked product. An example is given in Figure 1. Here, the effectiveness of cross-linking is compared for triallylamine and 1,3-butyleneglycol diacrylate. The amount of cross-linker has been normalized in each case for the functionality of the cross-linker, using the normalizing factor (1-2/f) where f is the functionality of the cross-linker. In Figure 1, the functionality of 1,3-butyleneglycol diacrylate is 4 and the functionality of triallylamine is 6. Several times more triallylamine is needed than diacrylate to achieve the same swelling ratio.

Specific Preparative Techniques

Graft Copolymerizations. Water soluble polymers such as starch and poly(vinyl alcohol) are grafted into superabsorbents in order to modify the properties. Certain processes benefit from increased viscosity of the monomer solution, and the water soluble, graft substrates can serve this purpose (*38,39*). Historically, the "superslurper" absorbents were made from acrylonitrile grafts to starch, and the currently used acrylic acid grafts to starch appear to have developed from the earlier work. Special initiators are useful to increase grafting efficiency with polymers containing hydroxyl sites (*13*). Metal ions, such as cerium, complex with the hydroxyl sites and serve as a locus for the nascent free-radical formed by oxidation of the substrate. For example, starch has been reacted with partially neutralized acrylic acid and a diacrylate ester cross-linking agent, using Ce^{4+} as a free radical initiator. The gel-like reaction product was dried and pulverized (*40*). Graft copolymers can also be prepared using an inverse suspension process (*41*).

Suspension Polymerization. In suspension polymerization processes, small droplets of aqueous monomer solution are dispersed into a second phase, usually aromatic or aliphatic hydrocarbon, prior to polymerization. Typically, the extent of neutralization is higher in suspension polymerizations because of the solubility of acrylic acid in the hydrocarbon phase, which leads to reaction and handling difficulties. Free radical polymerization of the monomers is then conducted in a manner similar to that described above for solution polymerization.

   Process Benefits. Process benefits of suspension polymerization include improved mixing of reagents during and after polymerization, no grinding of the product necessary to obtain the desired particle size and particle size distribution, and simpler removal of the heat of polymerization due to the fluidity of the reaction medium. The low viscosity of the suspension aids agitation, pumping, and heat removal from the polymerization.

   Control of Particle Size. Even though particle size of the product can be controlled, suspensions of relatively large particles with a narrow particle size distribution are difficult to stabilize in the early stages of polymerization. Examples of

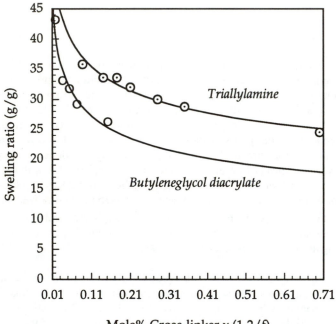

Figure 1. Dependence of the swelling ratio on the cross-link density and the functionality f of the cross-linker. The two cross-linkers differ in efficiency.

dispersing agents and the particle sizes produced when they are used are shown in Table III. Sorbitan monostearate as dispersing agent yielded particles with diameters in the range of 0.01–0.12 mm (*42*). Sorbitan monolaurate (*43*) yielded particle diameters in the 0.1–0.5 mm range, but more reactor waste was realized due to poor suspension quality. When saccharose mono-, di- or tristearates were used, the HLB of the specific agent used helped control the size of the polymeric particle formed (*44*). HLB of 2-6 yielded 100-500 micron beads, HLB of 6-16 yielded 100-500 micron granules and HLB of less than 2 gave lumps.

Combination suspending aids are also used. Finely ground bentonite or kaolin clays, modified by reaction with fatty amines, are used to achieve particle sizes of 5 micron to 1 mm in diameter (*45*). A combination of hydrophobic silica and a copolymer of acrylic acid and laurylmethacrylate as the suspending agents also yielded larger particles (*46*).

Control of Particle Shape. The shape of the particles produced in suspension processes can be affected by the viscosity of the monomer phase. Elongated SAP particles, with a narrower particle size distribution, can be made by adjusting the viscosity of the aqueous phase to above 5,000 cP with water-soluble thickeners (*39*). Viscosities up to 5000 cP yield spherical particle shapes only (and 100-600 micron diameters), whereas viscosities in the range 5000 to 1 million cP yield elliptical particles having lengths to 20 times their width (100-10000 micron). Mixtures of spherical and elongated particles were produced when the viscosity was between 5000 and 20,000 cP.

Table III. Effect of Suspending Agent on Particle Size

Suspending Agent	*Particle size produced*
1. Sorbitan monostearate	10-120 mm
2. Sorbitan monolaurate	150-500 mm
3. Sucrose / fatty acid esters	
HLB < 2	irregular lumps
HLB = 2-6	100-500 mm beads
HLB = 6-16	100-500 mm agglomerates
4. Ethyl cellulose	100-350 mm
5. Fatty amine modified clay	5 mm-1 mm
6. Silica/ modified polylaurylmethacrylate	150-1000 mm

Drying. An integral part of any process for superabsorbents prepared in an aqueous phase is drying of the gel. The essential principles involved were described some time ago (*47*). Polymers made in solution typically are dried after the gel particle size has been reduced, in order to speed drying through an increase in surface area. Continuous, hot air dryers frequently are used (*14*). Drum dryers (*2*) and continuous, screw dryers (*48*) have also been described. Polymers prepared in hydrocarbon suspensions are dried by azeotropic removal of the water from the suspension, followed by filtration of the dried particles from the hydrocarbon solvent. In a process designed to avoid a separate drying step, potassium acrylate is polymerized as a 70% solution of monomer in water (*11*). Polymerization commenced at 80°C and the water was evaporated from the polymer by the evolved heat of polymerization, yielding a substantially dry, porous product.

Surface Cross-linking. The polymer may be further cross-linked at the surface of the particles to alter the absorption rate of the product (*49, 50*). Surface coating or surface cross-linking of superabsorbent particles improves the wetting of the particles, prevents gel-blocking, and thereby improves the rate of water absorption. The cross-linking is by polyvalent metal ion salts, multi-functional organic cross-linkers, free-radical initiators or monomer coatings that are then polymerized. The reagents are added directly to the dry powder, or as solutions in water and organic solvents (*8*), or as dispersions in hydrocarbon solvents (*51*), or as blends with silicas or clays (*52*). Often a heating step is necessary after the coating step to promote the reaction of the cross-linker with the polymer surface.

Conclusion

The important physical properties of superabsorbent polyacrylates are dependent on the precise structure of the polymer network. Of key importance for use in personal care applications are the swelling capacity and the modulus of the swollen gel. The monomer and cross-linker concentrations, the initiator type and concentrations, the relative reactivities and efficiencies of the monomers and control of the reaction temperature are all significant.

Commercially, superabsorbent polymers are prepared from acrylic acid and sodium acrylate by solution cross-linking copolymerization (with or without a graft substrate), or suspension cross-linking copolymerization. Cross-linking copolymerization in solution suffers from the necessity of handling a rubbery reaction product. Suspension techniques simplify the handling of product but introduce the complexities of a hydrocarbon phase and suspending agents. Given the trade-offs for each, it should come as no surprise that both methods are practiced commercially.

Literature Cited
1. Kimura, K.; Hatsuda, T.; Nagasuna, K. EP 0450924 (1991).
2. Brandt, K. A.; Goldman, S. A.; Inglin, T. A. US 4,654,039 (1987); US Reissue. 32,649 (1988).
3. Kellenberger, S. R., European Patent Appl. EP 0339461 (1989).
4. Borchard, W., *Progr. Colloid Polym. Sci.*, **1991**,*86*, 84-91.
5. Geissler, E.; Hecht, A.; Horkay, F.; Zrinyi, M. *Macromolecules*, **1988**, *21*, 2594-2599.
6. Blanchette, A. R. Liquid Chromatographic Methods for Determining Extractable Monomeric Acrylate in Polyacrylate Absorbents, Institute for Polyacrylate Absorbents, Washington, DC, (1987).

7.   Cutie, S. S. *Analytica Chimica Acta*, **1992**, *260*, 13.
8.   Yoshinaga, K.; Nakamura, T.; Itoh, K. US 5,185,413 (1993).
9.   Flory, P. J. *Principles of Polymer Chemistry*, Cornell University Press, Ithaca, NY, 1953.
10. Siddall, J. H.; Johnson, T. C. US 4,833,222 (1989).
11. Takeda, H.; Taniguchi, Y. US 4,525,527 1985.
12. Chen, C-Y; Chen, C-S; Kuo, J-F *Polymer*, **1987**, *28*, 1396-1402.
13. Bazuaye, A.; Okieimen, F. E.; Said, O. B. *J. Polym. Sci.: Polymer Letters*, **1989**, *27*, 433-436.
14. Irie, Y.; Iwasaki, K.; Hatsuda, T.; Kimura, K.; Harada, N.; Ishizaki, K.; Shimomura, T.; Fujiwara, T. US 4,920,202 (1990).
15. Yano, K.; Kajikawa, K.; Nagasuna, K.; Irie, Y. European patent application EP 559476 (1993).
16. Tsubakimoto, T.; Shimomura, T.; Irie, Y.; Masuda, Y. US 4,286,082 (1981).

17. Chambers, D. R.; Fowler Jr., H. H.; Fujiura, Y.; Masuda, F. US 5,145,906 (1992).
18. Nowakowsky, B.; Beck, J.; Hartmann, H.; Vamvakaris, C. US 4,873,299 (1989).
19. Cizek,A.; Rice, H. L.; Thaemar, M. O. US 3,665,035 (1972).
20. Blay, J. A.; Shahidi, I. K. US 3,904,685 (1975).
21. Goretta, L. A.; Otremba, R. R. US 4,143,222 (1979).
22. Lenka, S.; Nayak, P. L. *J. Poly. Sci.:Part A Poly. Chem.*, **1987**, *25*, 1563.
23. Goretta, L. A.; Otremba, R. R. US 4,307,215 (1981).
24. Muenster, A.; Rohmann, M. US 4,301,266 (1981).
25. Furuno, A.; Inukai, K.; Ogawa, Y. US 4,514,551 (1985).
26. Cramm, J.; Bailey, K. US 4,698,404 (1987).
27. Yoshinaga, K.; Nakamura, T.; Itoh, K. EP 0398653 (1990).
28. Nagasuna, K.; Kadonaga, K.; Kimura, K.; Shimomura, T. EP 372981 (1989).
29. Hosokawa, Y.; Kobayashi, T. US 4,812,486 (1989).
30. Aoki, S; Yamasaki, H. US 4,093,776 (1978).
31. Ito, H.; Shimizu, S.; Suzuki, S. *J.Chem. Soc. Japan, Ind. Chem. Sect.*, **1955**, *58(3)*, 194-196.
32. Kabanov, V. A.; Topchiev, D. A.; Karaputadze, T. M. *J. Polym. Sci.: Symposia*, **1973**, *42*, 173-183.
33. Kabanov, V. A.; Topchiev, D. A.; Karaputadze, T. M.; Mkrtchian, L. A. *Eur. Polym. J.*, **1975**, *11*, 153-159.
34. Manickam, S. P.; Venkatarao, K.; Subbaratnam, N. R. *Eur. Polym. J.*, **1979**, *15*, 483-487.
35. Greenley, R. Z. *J. Macrom. Sci.-Chem.*, **1975**, *A9*, 505.
36. Buchholz, F. L. In *Absorbent Polymer Technology*, Brannon-Peppas, L.; Harland, R. S., Eds.; Elsevier Science Publishers, Amsterdam, 1990.
37. Butler, G. B. *Acc. Chem. Res.*, **1982**, *15*, 370-378.
38. Chmelir, M.; Pauen, J. US 4,857,610 (1989).
39. Nagasuna, K.; Namba, T.; Miyake, K.; Kimura, K.; Shimomura, T. US 4,973,632 (1990).

40. Masuda, F.; Nishida, K.; Nakamura, A. US 4,076,663 (1978).
41. Heidel, K. US 4,777,232 (1988).
42. Aoki, S.; Yamasaki, H. US 4,093,776 (1978).
43. Obayashi, S.; Nakamura, M.; Fujiki, K.;Yamamoto, T. US 4,340,706 (1982).
44. Nakamura, M. US 4,683,274 (1978).
45. Heide, W.; Hartmann, H.;Vamvakaris, C. US 4,739,009 (1988).
46. Stanley Jr., F. W.; Lamphere, J. C.; Chonde, Y. US 4,708,997 (1987).
47. Gehrmann, D.; Kast, W. *Proc. Int. Symp. Drying, 1st Ed.,* Mujumdar, A. S.,
    Ed.; Scientific Press, Princeton, NJ, 1978, pp. 239-246.
48. JP 02240112 (1990) to Chemie Linz A.-G.
49. Tsubakimoto, T.; Shimomura, T.; Irie, Y. GB 2,119,384-A (1983).
50. Yamasaki, H.; Kobayashi, T.; Sumida, Y. US 4,497,930 (1985).
51. Nagasuna, K.; Kadonaga, K.; Kimura, K.; Shimomura, T. EP 0317106, (1988).
52. Mikita, M.; Tanioku, S. US 4,587,308 (1986).

RECEIVED April 26, 1994

PROPERTIES

Chapter 3

Dynamic Swelling of Ionic Networks

Application to Behavior of Dry Superabsorbents

Nicholas A. Peppas and Deepak Hariharan[1]

School of Chemical Engineering, Purdue University,
West Lafayette, IN 47907−1283

The dynamic swelling behavior of ionic networks used as superabsorbent materials was modelled using a non-Fickian equation for water transport and a set of Nernst-Planck equations to analyze the ion transport in and out of the superabsorbent. Solution of this model was achieved with numerical techniques to give the water uptake and other parameters as functions of pH, ionic strength, molecular size and the Sw number.

Superabsorbent materials are used in a wide range of health care applications. They are especially important as materials for absorption of large amounts of fluids in diapers or incontinence products. Most commercially available superabsorbent materials are based on crosslinked polyacrylates (1).

Crosslinked polyacrylates are produced by polymerization of partially neutralized acrylic acid in the presence of a variety of tetra- or multifunctional crosslinking agents. Numerous patents (2) cover a wide range of temperatures and other reaction conditions under which such materials are produced.

The ensuing polyacrylates are dried and form microparticles, usually of irregular shape, that can be stored for a long period of time. When in contact with water, saline or any other fluid simulating urine, these particles have a tendency to swell rather fast and absorb the fluid. Typical superabsorbent materials are swollen within 5 to 10 sec, although equilibrium swelling may be attained much later.

Although numerous techniques have been proposed for the study of the swelling process (see for example references (3) and (4)), there has not been any attempt to analyze the dynamic swelling behavior of these polyacrylates, especially in the complex fluids that they will contact during their application.

[1]Current address: National Starch and Chemical Company, 10 Finderne Avenue, Bridgewater, NJ 08807

0097−6156/94/0573−0040$08.00/0
© 1994 American Chemical Society

Superabsorbent polyacrylates are ionic polymer networks containing ionizable pendant groups. The ability of these networks to swell many times their original weight upon ionization of their pendant groups renders them useful in many other interesting applications like in the development of biosensors and drug delivery devices. The rate of swelling and the equilibrium swelling ratio of these networks are the primary factors which affect their selection in various applications.

The transport of water into these polymer networks has been traditionally described in terms of Fick's second law. Unfortunately, most of the models developed neglect the macromolecular relaxation of the polymer during the transport process (5). These are serious omissions because during the transport of water, the glassy polymer network is transformed to a rubbery state with an associated stress relaxation in the network. Also, the water diffusion coefficient in the glassy region of the polymer is orders of magnitude smaller than the diffusion coefficient in the rubbery region.

Very few modelling efforts have been reported to analyze the dynamic swelling behavior of ionic networks. Kou and Amidon (6) used concentration-dependent diffusion coefficients to describe the water transport in crosslinked poly(hydroxyethyl methacrylate -co- methacrylic acid) copolymers. However, the associated swelling front velocity in the polymer was expressed in terms of experimental data without any physical foundation. Grodzinsky and coworkers (7) developed a model using Nernst-Planck equations to describe the transport of ionic species into such ionic networks. However, the polymer was assumed to be rubbery throughout the transport process. Hence the water diffusion coefficient was assumed to be constant.

A useful guide to predict the mode of transport in polymers was provided by Vrentas and Duda (8) in the form of the diffusional Deborah number, De, defined as the ratio of the characteristic relaxation time to the characteristic diffusion time. In general, when De is very much greater than or lower than one, normal Fickian diffusion is observed, whereas when De is close to 1, anomalous transport is observed. A particular case of non-Fickian transport, known as Case II transport, exists in which the swelling front moves at a constant velocity. The swollen region is assumed to be in equilibrium state of swelling, although transport is not the same as that of an unstressed gel due to constraints imposed by the glassy core. The non-Fickian transport has been described in many different ways by various researchers (9,10).

In this contribution, we developed models to describe water and ion transport in superabsorbents as a combination of Fickian diffusion and Case II transport. The model described below can be applied to the case of transport of simulated physiological solutions into an ionic superabsorbent polymeric materials and the effect of various parameters on the transport can be studied.

Model

Consider the case of a thin superabsorbent polymer layer (as in Figure 1) placed in water containing ions. The following subscripts are used in all further notation: 1 : water; 2 : polymer; 3 through n : ions.

The transport of water in a thin slab of an initially dry, hydrophilic polymeric network is described by a combination of Fick's second law and a pseudoconvective contribution to transport as indicated in Equation (1)

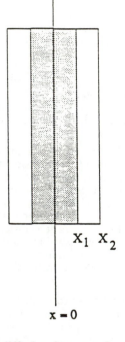

x_1 x_2

$x = 0$

Figure 1.   Schematic of diffusion into a polymer disk. The glassy region
                   extends from $x=0$ to $x=x_1$ and the rubbery region extends from
                   $x=x_1$ to $x=x_2$.

$$\frac{\partial c_1}{\partial t} = \frac{\partial}{\partial x}\left[D_1 \frac{\partial c_1}{\partial x} \right] - v \frac{\partial c_1}{\partial x} \tag{1}$$

Here, c_1 is the concentration of water, D_1 is the diffusion coefficient of water into the network and v is the pseudoconvective velocity as described by Frisch and coworkers (9) and later used by Peppas and coworkers (11).

Urine and physiological solutions are weak electrolytes and hence the transport of ions can be described by Nernst-Planck pseudobinary relations. In our model, the dissociation of electrolytes is assumed to be fast compared to transport. The transport of ions is described by a Nernst-Planck type pseudobinary equation.

$$\frac{\partial c_i}{\partial t} = \frac{\partial}{\partial x}\left[D_{im} \frac{\partial c_i}{\partial x} \right] + \frac{\partial}{\partial x}\left[\frac{c\, v_i\, D_{im} F}{RT} \frac{\partial \phi}{\partial x} \right] \tag{2}$$

Here, c_i is the concentration of ionic species i (3 through n), v_i is the charge on species i, F is the Faraday constant, ϕ is the electrostatic potential, R is the universal gas constant and T is the temperature. The requirement of electroneutrality at each point inside the polymer and the requirement of no current flow is also met.

Typical dry superabsorbent polymers contain an amount of extractables (usually monomer or oligomers and unreacted crosslinking agent), which are incorporated in the main network structure and may be of the order of 1-4 wt%. These extractables are released during the dynamic swelling process. They can be lumped together here as a "solute" and their transport can be analyzed using the same equation (2), where the subscript s indicates the extractables and is one of the indices i.

The diffusion of each of the mobile species is water-concentration dependent according to Fujita relation (12) as described by equation (3), where, D_{io} is the diffusion coefficient of species i at equilibrium and b_i is a constant characteristic of the polymer/diffusant system. This equation is based on the free volume theory.

$$D_{im} = D_{io} \exp\left[-b_i \left[1 - \frac{c_1}{c_1^*} \right] \right] \tag{3}$$

For a thin polymer film, initially in the dry, glassy state, the water and ion concentrations are zero so that

$$t = 0 \qquad -L_0 < x < L_0 \qquad c_i = 0 \ (i = 1,4,5...n) \tag{4}$$

The concentrations of water and ions at the polymer/water interface are assumed at equilibrium at all times and hence,

$$t \geq 0 \qquad x = \pm L_t \qquad c_i = c_{i,e}q \ (i = 1,4,5...n) \tag{5}$$

Here, $c_{i,eq}$ is the concentration of species at equilibrium. Also, the flux at the axis of the polymer is zero at all times and hence,

$$t \geq 0 \qquad x = 0 \qquad D_{im} \frac{\partial c_i}{\partial x} = 0 \ (i = 1,3,4.....n) \qquad (6)$$

The electrostatic potential at the polymer/water interface is zero at all times and the gradient of electrostatic potential is zero at the axis of the polymer. Hence,

$$t \geq 0 \qquad x = \pm L_t \qquad \phi = 0 \qquad (7)$$

and

$$t \geq 0 \qquad x = 0 \qquad \frac{\partial \phi}{\partial x} = 0 \qquad (8)$$

The transport of the superabsorbent extractables (subscript s) is further expressed in terms of the dimensionless Swelling Interface number, Sw, defined as follows.

$$Sw = \frac{v \, L_0}{D_{s_0}} \qquad (9)$$

Here, v is the pseudoconvective velocity of equation (1) and D_{s_0} is the preexponential term of the diffusion coefficient of the extractables as presented by equation (3). Thus, the Swelling Interface number indicates how fast the extractables will diffuse out of the superabsorbed material as a function of water swelling.

The swelling of the polymer can be described by knowledge of the water in the polymer network and assuming that the swelling of the polymer is ideal. It is also assumed that the polymer is initially glassy and is converted to rubbery as the water concentration increases inside the polymer. The transition of the polymer from a glassy to a rubbery state is characterized by a threshold concentration, c_w^+, which is the minimum local concentration of water required for the glassy state to be transformed into a rubbery state.

When the polymer is partially glassy, the sample is prevented from swelling in the axial direction and hence the thickness of the sample at time t, is given by

$$\frac{L_t}{L_0} = 1 + \frac{\int_{L_g}^{L_t} c_w dx}{\int_{L_g}^{L_t} c_{w,eq} dx} \left\{ \frac{1}{\upsilon_{2,eq}} - 1 \right\} \qquad (10)$$

where, $c_{w,eq}$ is the concentration of water in the swollen polymer at equilibrium and $v_{2,eq}$ is the volume fraction of polymer in the swollen polymer at equilibrium. It is assumed that the ionic species concentration is sufficiently low so that the total volume is predominantly the volume of water and polymer. Once, the polymer is completely transformed to rubbery state, the expression must be changed to

$$
\frac{L_t}{L_0} = \left[1 + \frac{\int_{L_g}^{L_t} c_w \, dx}{\int_{L_g}^{L_t} c_{w,eq} \, dx} \left\{ \frac{1}{v_{2,eq}} - 1 \right\} \right]^{1/3}
\tag{11}
$$

The concentration of various species at equilibrium is obtained using equilibrium thermodynamics as described in detail in another paper (*13*).

Results and Discussion

The system of equations presented above was solved by a finite difference technique. The water, ions and extractables transport equations were transformed into finite difference equations and solved by a forward time, centered space approximation. At each point of time, the concentrations of various species were calculated from the concentration at the previous time step. After the concentrations were calculated, the polymer volume was adjusted according to the mass balance equation. Once the new polymer volume was known, it was again divided into the same number of grids and the concentrations at each grid point were calculated assuming that the concentrations were linear in each grid. The electrostatic potential at each point was calculated by assuming electroneutrality at each point inside the polymer and hence equation (2) was used at each point in the polymer and the electrostatic potential was calculated using center space approximation.

Water uptake in initially dry superabsorbents was significantly influenced by the Swelling Interface number. The water uptake is shown in Figure 2 as a function of the dimensionless time τ for different values of the Swelling Interface number, i.e., Sw = 0.5, 1.0 and 5.0. Increase in the Swelling Interface number implied an increase of the pseudo-convective velocity which in turn implied the increase of the velocity of the glassy-rubbery front. This gave a polymer sample which was relaxing at a faster rate and hence the model should predict an increase in the uptake of water with increase in the Swelling Interface number. As the uptake increased to about 90% of the equilibrium value a discontinuity is observed in its slope. This is due to the fact that the glassy-rubber fronts have reached the center of the sample. Henceforth, the sample is completely rubbery and capable of swelling in all directions.

As it has been mentioned earlier, the superabsorbent is converted from glassy to rubbery state as the concentration of water in the polymer increases. As the glassy core in the polymer shrinks and ultimately disappears the polymer undergoes a readjustment of its volume. This could be further clarified when the sample thickness was plotted as a function of dimensionless time τ for Sw = 0.5, 1.0 and 5.0. The results of Figure 3 indicated that when the water concentration in the center of the sample became greater than the threshold concentration then the sample shrunk in the direction of diffusion. Hereafter, the sample was capable of swelling in all directions because there is no restriction for swelling in the x direction and hence it readjusted its volume.

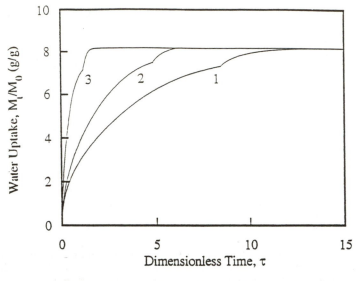

Figure 2.   Water uptake in a polymer disk as a function of dimensionless time,
                   τ for different Swelling Interface numbers. Curve 1: Sw = 0.5,
                   Curve 2: Sw = 1.0 and Curve 3: Sw = 5.0. Gels simulated here had
                   $\overline{M}_c$ = 6,000, concentration of ionizable groups, c_p = 1.0 mol/L, pK_b
                   = 4.7, polymer-solvent interaction parameter, χ = 0.47, ionic
                   strength of citrate-phosphate-borate buffer solution = 0.1M and pH
                   of buffer solution = 4.0.

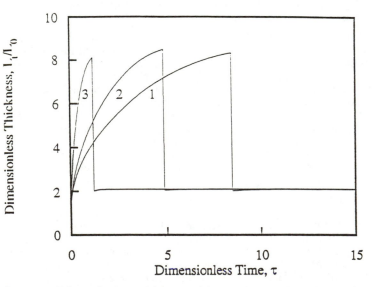

Figure 3.   Dimensionless thickness of a polymer disk as a function of
                   dimensionless time, τ for different Swelling Interface numbers.
                   Curve 1: Sw = 0.5, Curve 2: Sw = 1.0 and Curve 3: Sw - 5.0. Gels
                   simulated here had the same parameters as in Figure 2.

As it has been already mentioned, the properties of ionic polymers are dependent on the concentration of their ionic groups. The variation of the total concentration of ionized groups as a function of time at different pH values of a citrate-phosphate-borate buffer is presented in Figure 4. It can be seen that the rate of ionized groups concentration increased with a decrease in pH. The final concentration of total ionized groups was the same at pH 2 and 6 because under these conditions all the groups are ionized. However, at pH value of 10 only a fraction of the groups are ionized and hence we see a reduction in the total concentration of ionized groups.

The concentration of pendant ionic groups also plays a role in the water uptake by the copolymeric sample. It is important to study the effect of initial ionic group concentration in the sample on the water uptake of the sample. The effect of ionic group concentration on water uptake is depicted in Figure 5 for ionic group concentrations of 0.1 mol/L and 1.0 mol/L. The water uptake increased with an increase in concentration of ionic groups, because of an increase in repulsion between the ionized groups which increased the swelling of the polymer sample.

Finally, the effect of pH on the water uptake is presented in Figure 6. As the pH decreased there was an increased uptake which is due to the increased ionization of the groups in the sample.

These results indicate that it is possible to model the swelling behavior of a superabsorbent polymer taking in consideration the various physicochemical parameters of the problems and the various ions present.

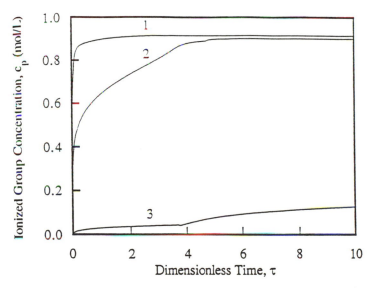

Figure 4.   Ionized Group concentration in the polymer disk as a function of dimensionless time, τ for different pH of external buffer solution. Curve 1: pH 2, Curve 2: pH 6 and Curve 3: pH 10. Gels simulated here had $\overline{M}_c$ = 6,000, Swelling Interface number, Sw = 1.0, concentration of ionizable groups, c_p = 1.0 mol/L, pK_b = 4.7, polymer-solvent interaction parameter, χ = 0.47, ionic strength of citrate-phosphate-borate buffer solution = 0.1M.

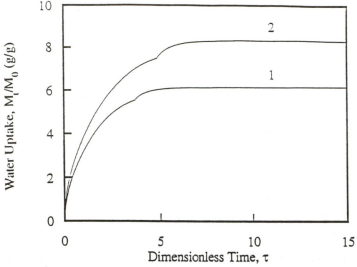

Figure 5.   Water uptake in a polymer disk as a function of dimensionless time, τ for two initial concentrations of ionizable groups. Curve 1: c_p = 0.1 mol/L and Curve 2: c_p = 1.0 mol/L. Gels simulated here had $\overline{M}_c$ = 6,000, Swelling Interface number, Sw = 1.0, pK_b = 4.7, polymer-solvent interaction parameter, χ = 0.47, ionic strength of citrate-phosphate-borate buffer solution = 0.1M and pH of buffer solution = 4.0.

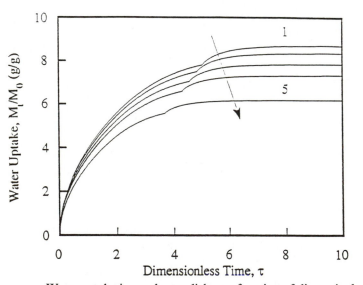

Figure 6.   Water uptake in a polymer disk as a function of dimensionless time, τ for different pH of the external solution. Curve 1: pH = 2, Curve 2: pH = 4, Curve 3: pH = 6, Curve 4: pH = 8 and Curve 5: pH = 10. Gels simulated here had $\overline{M}_c$ = 6,000, Swelling Interface number, Sw - 1.0, pK_b = 4.7, polymer-solvent interaction parameter, χ = 0.47, ionic strength of citrate-phosphate-borate buffer solution = 0.1M.

References

1. Brannon-Peppas, L.; and Harland, R.S., *Superabsorbent Polymer Technology*, Elsevier, Amsterdam, **1990.**

2. Buchholz, F., in *Superabsorbent Polymer Technology*, Brannon-Peppas, L. and Harland, R.S., eds., Elsevier, Amsterdam, **1990**, pp 27-44.

3. Dubrovskii, S.A; Afanas'eva, M.V.; Lagutina, M.A.; Kazanskii, K.S., *Vysokomol. Soyed.*, **1990**, *A32*, 165-170.

4. Khare, A.R.; Peppas, N.A.; Massimo, G.; Colombo, P., *J. Controlled Release*, **1992**, *22*, 239-244.

5. Rossi, G.; Mazich, K.A., *Phys. Rev. E.*, **1993**, *48*, 1182-1185.

6. Kou, J.; Amidon, G., *Proceed. Intern. Symp. Control. Rel. Bioact. Mater.*, **1987**, *14*, 79-80.

7. Crimshaw, P.E.; Grodzinsky, A.J.; Yarmush, M.L.; Yarmush, D.M., *Chem. Eng. Sci.*, **1990**, *45*, 2917-2925.

8. Vrentas, J.; Duda, J.L., *J. Polym. Sci. Polym. Phys.*, **1977**, *15*, 441-457.

9. Frisch, H.L.; Wang, T.T.; Kwei, T.K., *J. Polym. Sci.*, **1969**, *A2*, 7, 879-888.

10. Lustig, S.R.; Caruthers, J.M.; Peppas, N.A., *Chem. Eng. Sci.*, **1992**, *12*, 3037-3057.

11. Peppas, N.A.; Sinclair, J.L., *Coll. Polym. Sci.*, *1983*, *261*, 404-408.

12. Fujita, H., *Adv. Polym. Sci.*, **1961**, *3*, 1-31.

13. Hariharan, D.; Peppas, N.A., *J. Membr. Sci.*, **1993**, *78*, 1-12.

RECEIVED August 9, 1994

Chapter 4

Mechanical Behavior of Swollen Networks

Burak Erman

Polymer Research Center and Tubitak Advanced Polymeric Materials Research Center, Bogazici University, Bebek 80815, Istanbul, Turkey

A network immersed in solvent is a semi-open thermodynamic system. The immersed network may absorb solvent if stretched and give out solvent if compressed. Measurements on networks of relatively low degrees of cross-linking immersed in solvents show that a large amount of solvent enters the network upon stretching even at relatively small extension ratios. Such large dimensional changes resulting from the movement of solvent into the network affects the elastic properties of the networks significantly. Elastic parameters, such as modulus of elasticity, shear modulus, bulk modulus and Poisson's ratio, depend strongly on the interaction of the polymer with the solvent. Values of elastic parameters are calculated in the present study for networks as a function of cross-link density, interaction parameter and degree of ionization.

Swelling of a network has a marked effect on its mechanical properties, which are measured either in air or in the immersed state. A network swollen and subject to a tensile force in air behaves as a closed system for which the volume fraction of solvent remains constant during deformation. A network immersed in solvent, on the other hand, is a semi-open thermodynamic system. It may absorb solvent if stretched and give out solvent if compressed. Measurements on *cis*-polyisoprene networks immersed in various solvents *(1)* show that a relatively large amount of solvent enters the network upon stretching even at relatively small extension ratios. For a moderately cross-linked polyisoprene network in n-hexane, for example, the volume fraction of polymer drops from 0.253 to 0.130 when stretched in the immersed state up to an extension ratio of 1.40. Diffusion coefficients associated with the process are in the order of 10^{-7}-10^{-8} cm^2/s *(1)*.

The equilibrium molecular theory of elasticity allows for an estimate of elastic coefficients and the degree of swelling of a network, in the isotropically swollen and

0097–6156/94/0573–0050$08.00/0

© 1994 American Chemical Society

the deformed state as a function of network constitution. This is true for a network that behaves as a closed system as well as a semi-open one. The mechanical behavior of a swollen network deformed in air has been studied extensively *(2)*. The mechanical behavior of a network exposed to the action of a solvent has not been studied in very much depth, however. The original work of Gee *(3)* and Flory and Tatara *(4)* are two examples. Recently, the subject has gained some importance, perhaps due to the technological importance of superabsorbent polymers and their use as semi-open systems. In the present study, the mechanical behavior of a network immersed in solvent will be reviewed and the modulus of elasticity, bulk modulus and Poisson's ratio of such networks will be calculated as a function of network constitution.

Free Energy of a Swollen Network

The free energy change ΔA resulting from swelling of an amorphous undeformed network with a solvent is given as the sum of three terms *(5, 6)*: first, the free energy of mixing ΔA_{mix}, second, the elastic free energy ΔA_{el} resulting from the dilation of the network and third, the contribution ΔA_i from the presence of ionic groups on the network chains:

$$\Delta A = \Delta A_{mix} + \Delta A_{el} + \Delta A_i \tag{1}$$

The last term in equation 1 is not commonly present in the classical theories of rubber elasticity. Its effect on the chemical potential will be approximated by a linear term below. The expression for ΔA_{mix} is readily obtained from the lattice theory of polymer solutions. The result is *(7)*

$$\begin{aligned} \Delta A_{mix} &= \Delta A_{comb} + A^R \\ &= RT(n_1 \ln v_1 + n_2 \ln v_2 + \bar{\chi} \, n_1 \, v_2) \end{aligned} \tag{2}$$

Here, ΔA_{comb} and A^R denote the combinatorial and the residual free energy, n_1 and n_2 are mole numbers of solvent and polymer, and v_1 and v_2 are their volume fractions. In equation 2, $\Delta A_{comb} = kT(n_1 \ln v_1 + n_2 \ln v_2)$ is obtained from the generalized ideal mixing law *(8)*. The residual free energy is in general a function of concentration $A^R = \bar{\chi} \, n_1 \, v_2$. The quantity $\bar{\chi}$ is the dimensionless interaction parameter for the solvent-polymer system, and is a function of concentration. The quantity $kT\bar{\chi}$ is equal to the difference in energy of a solvent molecule immersed in pure polymer compared with a solvent molcule in the pure state.

The elastic free energy in equation 1 has been written for different molecular models *(2)* such as the constrained-junction, constrained chain and slip-link models. The present discussion will be based on the constrained-junction model *(9,10)* which is the simplest form of the various entanglement models. Accordingly, the elastic free energy is

$$\Delta A_{el} = \frac{1}{2} \xi kT \sum_t \{ (\lambda_t^2 - 1) + \tag{3}$$

$$(\mu / \xi) [B_t + D_t - \ln (B_t + 1) - \ln (D_t + 1)]\}$$

Here, ξ is the cycle rank of the network, i.e., the number of chains that have to be cut to reduce the network to a tree. Defined in this manner, ξ is independent of the degree of perfection of a network and is a universal measure of elastic activity. For slightly cross-linked networks which are highly imperfect by construction, ξ is the proper structural parameter. μ is the number of elastically active junctions. λ_t is the ratio of the final length along the t'th principal coordinate direction to the length in the undeformed state, and

$$B_t = \kappa^2 (\lambda_t^2 - 1) (\lambda_t^2 + \kappa)^{-2} \tag{4}$$

$$D_t = \kappa \lambda_t^2 (\lambda_t^2 - 1) (\lambda_t^2 + \kappa)^{-2} = \lambda_t^2 \kappa^{-1} B_t \tag{5}$$

The parameter κ that appears in equations 4 and 5 is the entanglement parameter, which is zero for a phantom network and infinity for an affine network. For a real network, adopting a finite value of κ leads to good agreement with experimentally determined values of the reduced force.

The components λ_t of the deformation may be decomposed into two parts as

$$\lambda = (v_{2c}/v_2)^{1/3} \alpha = (V/V^0)^{1/3} \alpha \tag{6}$$

Here, v_{2c} is the volume fraction of polymer present during network formation. λ is the diagonal tensorial representation for λ_t. The definition of α follows from equation 6. The nonzero terms α_t along its diagonal are the extension ratios defined as the ratio of the final length along the t'th principal direction to the swollen, undistorted length. Defined in this way, α_t becomes a measure of distortion of the network. For swelling without distortion, α equates to the third-order unit tensor $\mathbf{E}$, and λ has three equal diagonal components, and

$$\lambda = (v_{2c}/v_2)^{1/3} \mathbf{E} \tag{7}$$

For uniaxial deformation along the x-direction,

$$\lambda_1 = \lambda.$$

$$\lambda_2 = \lambda_3 = [v_{2c}/(v_2\lambda)]^{1/2} = (v_{2c}/v_2)^{1/2}\alpha^{-1/2}$$

(8)

Once the free energy of the system is defined, several useful thermodynamic and mechanical relationships may be obtained for the swollen network as will be discussed in the following sections.

The Reduced Solvent Chemical Potential for an Isotropically Swollen Network

The chemical potential, $\Delta\mu_1 = \mu_1 - \mu_1^{0}$, of solvent in the swollen network is obtained by differentiating equation 1 with respect to the number of solvent molecules at fixed temperature T, and pressure p *(5)*. A more convenient quantity to use is the reduced chemical potential, $\Delta\tilde{\mu}_1 = \Delta\mu_1/RT$:

$$\Delta\tilde{\mu}_1 = \ln a_1 = (RT)^{-1}(\frac{\partial\Delta A}{\partial n_1})_{T,p}$$
$$= (RT)^{-1}\left[(\frac{\partial\Delta A_{mix}}{\partial n_1})_{T,p} + (\frac{\partial\Delta A_{el}}{\partial n_1})_{T,p} (\frac{\partial\Delta A_i}{\partial n_1})_{T,p}\right]$$
$$= (\Delta\tilde{\mu}_1)_{mix} + (\Delta\tilde{\mu}_1)_{el} + (\Delta\tilde{\mu}_1)_i$$

(9)

The quantity a_1 in the first line of equation 9 represents the activity of the solvent in the network. The reduced chemical potential due to the mixing term is obtained from equation 2 as

$$(\Delta\tilde{\mu}_1)_{mix} = \ln(1 - v_2) + v_2 + \chi v_2^2$$

(10)

The term χ in equation 10 is related to the interaction parameter $\bar{\chi}$ by the following relations:

$$\chi \equiv \frac{1}{RTv_2^2} \frac{\partial A^R}{\partial n_1}$$
$$= \frac{\partial(v_2\bar{\chi})}{\partial v_2} - \frac{\partial\bar{\chi}}{\partial v_2}$$
$$= \chi_1 + \chi_2 v_2 + \chi_3 v_2^2 + ...$$

(11)

The second line of equation 11 is obtained by differentiating the third term of equation 2. The third line of equation 11 represents the Taylor series expansion for χ with only the first three terms shown. The parameters χ_1, χ_2 and χ_3 are now independent of concentration. The Equation of State approach (11), Random Phase Approximation (12), or the Polymer Reference Interaction Model (13) are some of the techniques by which the interaction parameter may be related to molecular variables of a polymer-solvent system.

The elastic part of the reduced chemical potential is obtained by using the chain rule

$$(\Delta\tilde{\mu}_1)_{el} = (\frac{\partial\Delta A_{el}}{\partial\lambda})_{T,p} (\frac{\partial\lambda}{\partial n_1})_T$$

$$= (\frac{\partial\Delta A_{el}}{\partial\lambda})_{T,p} \left[\frac{V_1}{3V^{2/3}V_0^{1/3}}\right]$$

(12)

The first factor on the right-hand side of equation 12 is obtained by differentiating equation 3 with respect to λ, after substituting $\lambda = \lambda_1 = \lambda_2 = \lambda_3$, and $\lambda^3 = V/V_0$, which corresponds to free swelling. The resulting expression is of the form

$$(\Delta\tilde{\mu}_1)_{el} = (\Delta\tilde{\mu}_1)_{el} = (\frac{\partial\Delta A_{el}/RT}{\partial\lambda})_{T,p} = \frac{3\xi}{N_A} [1 + \frac{\mu}{\xi} K(\lambda)]$$

(13)

The second factor on the right-hand side of equation 12 is obtained first by writing λ as

$$\lambda = (\frac{V}{V_0})^{1/3} = (\frac{n_1 V_1 + x V_1 n_2}{V_0})^{1/3}$$

(14)

where V_1 is the molar volume of solvent, and then differentiating with respect to n_1:

$$(\frac{\partial\lambda}{\partial n_1})_T = \frac{1}{3} \frac{V_1}{V^{2/3}V_0^{1/3}} = \frac{1}{3} \frac{V_1}{V_0 \lambda^2}$$

(15)

The reduced solvent chemical potential due to the elastic activity of the network follows from equations 13, 14 and 15 as

$$(\Delta\tilde{\mu}_1)_{el} = \frac{\beta}{\lambda} [1 + \frac{\mu}{\xi} K(\lambda)]$$

(16)

where $K(\lambda)$ is

$$K(\lambda) = \frac{B\dot{B}}{1 + B} + \frac{D\dot{D}}{1 + D} \tag{17}$$

$$\dot{B} = \frac{\partial B}{\partial \lambda^2} = B[(\lambda^2 - 1)^{-1} - 2(\lambda^2 + \kappa)^{-1}] \tag{18}$$

$$\dot{D} = \frac{\partial D}{\partial \lambda^2} = \kappa^{-1}(\lambda^2 \dot{B} + B) \tag{19}$$

$$\beta = \frac{V_1}{RT} \frac{\xi kT}{V_0} = \frac{v_{2c}}{x_c} \left(1 - \frac{2}{\phi}\right) \tag{20}$$

In equation 20, V_1 is the molar volume of solvent, ϕ is the functionality of junctions and x_c is the number of repeat units of the network chain, the volume of each being equal to the volume of a solvent molecule.

Finally, the chemical potential due to ionic groups along the network chains is written as *(8)*

$$(\Delta\tilde{\mu}_1)_i = -\frac{iv_2}{x_c} \tag{21}$$

Here, i is the number of ionic centers along each network chain.

The total reduced chemical potential of solvent in the swollen network is

$$\Delta\tilde{\mu}_1 = \ln(1 - v_2) + v_2 + \chi v_2^2 + \frac{\beta}{\lambda}\left[1 + \frac{\mu}{\xi}K(\lambda)\right] - \frac{iv_2}{x_c} = \ln a_1 \tag{22}$$

with $\lambda = (V/V_0)^{1/3}$ as described above. When the network is immersed in solvent, it swells by taking up solvent until an equilibrium is reached between entropic forces that encourage the solvent molecules to mix into the network, and the elastic forces of the stretched chains which tend to prevent further swelling. At conditions of equilibrium, the activity a_1 of solvent equates to unity and the solvent chemical potential equates to zero. The solution of equation 22 for v_2 then gives the equilibrium degree of swelling of the network.

Thermodynamics of a Uniaxially Stretched Network in Solvent

When a network at swelling equilibrium is stretched uniaxially in excess solvent to a fixed extension ratio λ, equilibrium will be re-established by the uptake of solvent of the network. Since the length is fixed, swelling takes place along the lateral directions. In this section we give the derivation of some mechanical material coefficients. For simplicity, we give the results for a network formed in the dry state, i.e., with $v_{2c} = 1$.

When the network is in equilibrium with the surrounding solvent at strain λ, the total volume will be V', the volume fraction of polymer will be v'_2, and the lateral extension ratio will be related (14) to the longitudinal one by

$$\lambda = \alpha \, v_2^{-1/3}$$

(23)

$$\lambda_2 = \lambda_3 = v_2^{1/6} (v')^{-1/2} \alpha^{-1/2}$$

The reduced solvent chemical potential at constant length is

$$\Delta\tilde{\mu}_1 = (RT)^{-1} \left(\frac{\partial \Delta A}{\partial n_1} \right)_{T, p, \lambda}$$

(24)

leading to

$$\Delta\tilde{\mu}_1 = \ln(1 - v'_2) + v'_2 + \chi \, v'^2_2 + \beta\lambda_2^2 v'_2 \left[1 + \frac{\mu}{\xi} K(\lambda_2) \right] - \frac{i v'_2}{x_c} = 0$$

(25)

Calculation of Elastic Coefficients for a Swollen, Immersed Network for Small Deformations

The tensile force f for simple deformation is given by

$$f = L_0^{-1} \left(\frac{\partial \Delta A_{el}}{\partial \lambda} \right)_{T, V, n_1}$$

(26)

where L_0 is the length of the isotropic undistorted sample in the direction of stretch at the reference volume V_0. Using equation 26 in equation 3 leads to the expression for the force. In the present paper, we derive the equations for small deformations, i.e., $\alpha \longrightarrow 1$. In this limit, the resulting expression is

$$f = \left\{ \frac{3\xi kT}{L_0} \left(\frac{v_2^0}{v_2} \right)^{1/3} [\, 1 + \frac{\mu}{\xi} \frac{\kappa^2(\kappa^2 + 1)}{(1 + \kappa)^4}] \right\} \varepsilon \tag{27}$$

Here, $\varepsilon = \alpha\text{-}1$, and v_2 is the volume fraction of polymer which equates to v_2' for the semi-open system, obtained as the solution of equation 25. We define the modulus of elasticity E of a swollen network as the force per unit swollen, undistorted area, A. Dividing both sides of equation 27 by A and relating to the unswollen dimensions leads to

$$E = \frac{3\xi kT}{V_0} \left(\frac{v_2'}{v_2^0} \right)^{1/3} [\, 1 + \frac{\mu}{\xi} \frac{\kappa^2(\kappa^2 + 1)}{(1 + \kappa)^4}] \tag{28}$$

The second term in the squared brackets results from the contribution of entanglements.

The Poisson's ratio, r, is defined as the negative of the ratio of lateral strain to longitudinal strain. In terms of deformation and swelling parameters it may be written as

$$r = - \frac{\dfrac{v_2^{1/3}}{(v_2'\lambda)^{1/2}} - 1}{\lambda v_2^{1/3} - 1} = - \frac{(\dfrac{v_2}{v_2'})^{1/2} - 1}{\alpha - 1} \tag{29}$$

In equation 29, the value of the equilibrium degree of swelling v_2 in the undistorted state is obtained from the solution of equation 22, and v_2' for the stretched state is obtained from equation 25.

From the mathematical theory of linear elasticity, the shear and bulk moduli are related to E and r by the following expressions *(15)*

$$G = \frac{E}{2(1 + r)} \tag{30}$$

$$K = \frac{E}{3(1 - 2r)} \tag{31}$$

Numerical examples

In this section, we present numerical values for r, E, G and K for networks whose molecular constitution is specified. As a special case, we adopt the phantom network model for which $\kappa=0$. The value of the dimensionless parameter β is chosen as 0.004. This results from the arbitrary choice of $V_1 = 100$ cm^3/mole and $\xi kT/V_0 = 0.1$ N/mm^2. For simplicity of presentation, we first present results for non-ionic networks followed by a treatment of ionic networks.

Non-ionic networks. The values of the modulus of elasticity, obtained from equation 28 with i = 0, subject to the solution of equation 25 are plotted in Figure 1, for three different values of χ. The Poisson ratio shows a strong dependence on the quality of solvent as shown in Figure 2. The $\varepsilon = 1$ limits of r are seen to be close to 1/2, 1/3 and 1/5 for $\chi = 1.0$, 0.5 and 0.0, respectively.

There have been various estimates of the Poisson ratio of gels in recent years, both theoretical and experimental. According to calculations based on the picture of a single chain in theta solvent, r is equal to zero (16). This is in disagreement with the results of the present calculations. According to scaling arguments, r = 1/4 in a good solvent (17). The present mean field Flory-Huggins type calculations give this limit as 1/5. Experiments performed on swollen networks are not conclusive in this matter. Osmotic compressibility data (18) on poly(vinyl acetate) gels in good solvents gave r = 0.2 which is in agreement with results of the present analysis. Previous approximate calculations (19) were reported as r $\approx$ 0.25. The present result, based on more rigorous calculations is the correct one. Light scattering and osmotic compressibility measurements (16) on polyacrylamide gels in good solvents led to values of r between 0.3 and 0.4.

The dependence of shear modulus G on the quality of solvent, obtained from equation 30 is shown in Figure 3. Similarly, the values of the bulk modulus, K, obtained from equation 31 are shown in Figure 4. It is to be noted from Figure 4 that the bulk compressibility is very strongly dependent on the quality of solvent. For poor solvents, K shows a very strong dependence on strain, being extremely large at small strains.

Ionic networks. Dependence of equilibrium degree of swelling on the number of ionic groups, obtained from the numerical solution of equation 22 for various values of i, is presented in Figure 5 for $\chi=0.0$ and 0.5. The strong effect of the presence of ionizable groups on swelling, also affects the elastic parameters. Values of the elastic modulus for $\chi=0.0$ are obtained from equation 25 for a network with $\kappa=0$ are shown in Figure 6. Values of v_2 that appear in equation 28 are obtained from Figure 5 and v_2' are obtained from equation 25. Values of the Poisson's ratio, obtained from equation 29 are also shown in Figure 6. The dependence of r on the number of ionic groups is further elaborated in Figure 7 where results of calculations are presented for a poor, theta and a good solvent. Insensitivity of r to i when $\chi=1.0$ and the non-linear strong dependence when $\chi=0.5$ is worth noting.

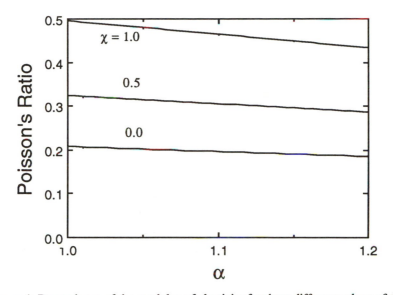

Figure 1. Dependence of the modulus of elasticity for three different values of the χ parameter.

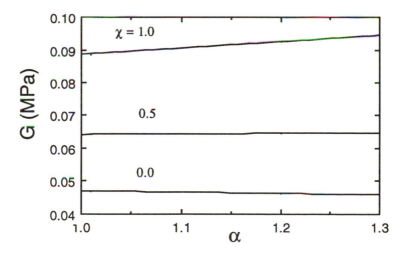

Figure 2. Dependence of Poisson's ratio for three different values of the χ parameter.

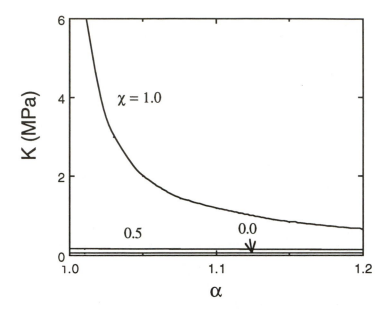

Figure 3. Dependence of shear modulus on strain for three different values of the χ parameter.

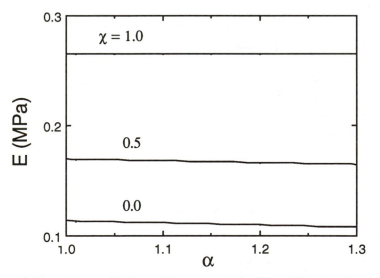

Figure 4. Dependence of bulk modulus on strain for three different values of the χ parameter.

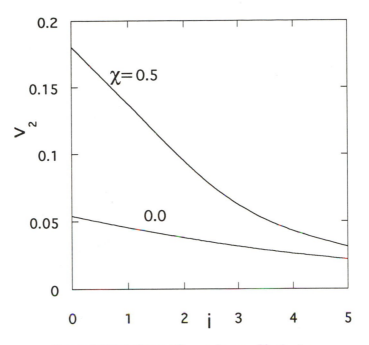

Figure 5. Dependence of v_2 on degree of ionization.

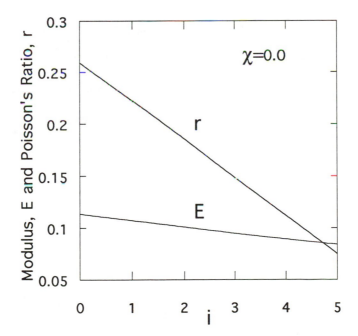

Figure 6. Dependence of the elastic modulus and Poisson's ratio on degree of ionization.

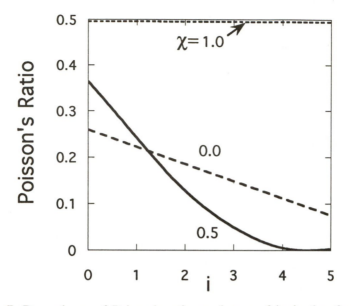

Figure 7. Dependence of Poisson's ratio on degree of ionization for three different values of the interaction parameter.

Discussion

The sensitive dependence of the elastic coefficients on both the degree of swelling and strain is of utmost importance and their effects are observed in various phenomena relating to the general mechanical-thermodynamic behavior of swollen networks. Calculations performed but not reported in the present paper show that the onset of volume phase transitions upon stretching an immersed network is sensitively dependent on the quality of solvent, and that extremely stringent conditions are required for the realization of spinodal as well as volume phase transitions. For strongly ionic networks, this may not be the case as preliminary calculations indicate. Work along this direction is in progress.

Acknowledgment. This work is partially supported by the Research Fund, Project Number 93P0084 of Bogazici University.

Literature Cited

(1)   Ozkul, M. H.; Onaran K.; Erman, B. *J. Polym. Sci.: Part B: Polym. Phys.*, **1990**, *28*, 1781.
(2)   Mark, J. E.; Erman, B. *Rubberlike Elasticity, A Molecular Primer*, Wiley-Interscience: New York, **1988**.
(3)   Gee, G. *Trans Faraday Soc..* **1946**, *42B*, 33.
(4)   Flory, P. J.; Tatara, Y. *J. Polym. Sci., Polym Phys. Ed.* **1975**, *13*, 683.

(5) Bahar, I.; Erman, B. *Macromolecules,* **1987**, *20*, 1696.
(6) Erman, B.; Flory, P. J. *Macromolecules,***1986**, 19, 2342
(7) Erman, B.; Flory, P. J., *Polym. Commun.,* **1984**, *25*, 132.
(8) Flory, P. J., *Principles of Polymer Chemistry,* Cornell University Press, New York, **1953**.
(9) Erman, B.; Flory, P. J., *Macromolecules*, **1982**, *15*, 800.
(10) Erman, B.; Flory, P. J., *Macromolecules*, **1982**, *15*, 806.
(11) Flory, P. J. *Discuss. Faraday Soc.,* **1970**, *49*, 7.
(12) de Gennes, P. G., *Scaling Concepts in Polymer Physics,* Cornell University Perss, New York, **1979**.
(13) Schweizer, K. S.; Curro, J. G., *Macromolecules*, **1988**, *21*, 3070.
(14) Erman, B.; Flory, P. J., *Macromolecules*, **1983**, *16*, 1607.
(15) Sokolnikoff, I. S. *Mathematical Theory of Elasticity*, McGraw Hill, **1956**.
(16) Geissler, E.; Hecht, A. M., *Macromolecules,* **1980,** *13*, 1276.
(17) Daoud, M.; de Gennes, P. G. *J. Phys. (Paris),***1977**, *38*, 85.
(18) Zrinyi, M.; Horkay, F., *J. Polym. Sci. Polym. Phys. Ed.*, **1982**, *20*, 815.
(19) Erman, B., Lecture presented in the Symposium on Advances in Superabsorbent Polymers, PMSE, ACS Chicago Meeting, August 1993.

RECEIVED April 26, 1994

Chapter 5

Ratio of Moduli of Polyelectrolyte Gels in Water With and Without Salt

Yong Li[1], Chunfang Li[2], and Zhibing Hu[2]

[1]Kimberley-Clark Corporation, Neenah, WI 54956
[2]Department of Physics, University of North Texas, Denton, TX 76203

The ratio of bulk modulus to shear modulus, K/G, of ionic polyacrylamide (PAAM) gels is studied as functions of degree of ionization and salt concentration. This ratio is directly related to the exponents characterizing the concentration dependence of bulk and shear modulus, respectively. It was found that the value of δ is sensitive to both the degree of neutralization and salt concentration. For non-ionic gel in pure water, $\delta=2.1$. As ionic groups are introduced into the network, the value of δ decreases quickly and reaches ~ 1.08. Once salt is added to the system, however, the *delta* value recovers and reaches 1.8. It was also observed that the ionization-salt space can be divided into four regions according to the behavior of δ.

From application point of view, polyelectrolyte gels offer some unique opportunities. They have much higher swelling capability for given stiffness, and typically are sensitive to a variety of conditions, including pH, solvent ionic strength (salt concentration), and external electric field [1, 2]. The large, discontinuous volume phase transition of these materials make them promising "intelligent materials" in various fields [3] (see also the chapter by Shimomura). Many authors have attempted to derive theories to describe the swelling behavior of these gels. Prud'homme and Yin recently developed a comprehensive polyelectrolyte gel swelling theory with both ion condensation and finite chain extension effects included [4]. Most of these work, however, were verified by swelling data and do not deal with the network mechanical properties in depth. More detailed study of mechanical properties of polyelectrolyte gels is thus needed to meet the challenges of both theoretical understanding and practical applications.

The mechanical properties of gels can be represented by two parameters, i.e., the bulk modulus K and shear modulus G. The bulk modulus is related to the derivative of gel osmotic pressure and therefore measures the ability of the net-

0097–6156/94/0573–0064$08.00/0
© 1994 American Chemical Society

work to swell (volume change) against external pressure. The shear modulus is a measure of the ability of the network to maintain its shape against external pressure. For example, the swelling ratio reduction of a gel when macromolecules are introduced in the solvent is related to its bulk modulus. Whereas the instantaneous deformation of a gel under external compressional force is related to its shear modulus. These two properties play a vital role in many phenomena, including the bending of polyelectrolyte gels in electric field [5], gel surface pattern formation [6], network swelling kinetics [7], gel volume phase transition [8], etc. In the case of superabsorbent polyacrylate used in personal care products (e.g., diapers), Masuda, Nagorski, and several other authors in other chapters have demonstrated that the performance of these products is directly related to superabsorbent mechanical properties measured by various methods, including Absorbency Under Load (AUL), shear modulus, re-absorbency under shear, and stability of modulus.

Recently we developed a simple method to measure the ratio of gel bulk modulus (K) to shear modulus (G) [9]. Using this technique, we studied acrylamide/sodium acrylate copolymer gels in water with and without sodium chloride (NaCl) salt. Since all highly ionized gels are hydrogels, throughout the rest of this chapter we will assume the basic solvent is water unless otherwise mentioned. The term "salt" refers to mono-monovalent salts throughout this paper.

Theoretical Background

Gels are 3-dimensional networks of crosslinked polymers in solvent. Associated with the crosslinking process is a free energy increase due to the reduction of entropy. In general, the free energy of a gel can be written as

$$F = F_1(\phi) + F_2(\alpha_x, \alpha_y, \alpha_z) \tag{1}$$

where ϕ is the polymer volume fraction, α_i's are the principal elongation ratios relative to the gel dimensions when it was made. For isotropic swelling, all α_i have the same value. The α's are related to the gel volume V and volume fraction ϕ,

$$\alpha_x \alpha_y \alpha_z = V/V_0 = \phi_0/\phi \tag{2}$$

where ϕ_0 is the polymer volume fraction at preparation, and V and V_0 are gel volumes at polymer volume fraction equal to ϕ and ϕ_0, respectively. The first term in Equation (1) is equivalent to the free energy of the corresponding uncrosslinked polymer solution (or, a gel with crosslinking "turned off"). This term is polymer concentration dependent only. The second term reflects the free energy of gels related with anisotropy and therefore is called elastic free energy. This term is a direct result of crosslinking. It is this free energy term that introduces an extra parameter, the shear modulus, into the gel system.

Following Flory-Huggins theory [10], the elastic free energy term can be written as

$$F_2 = B\left[\alpha_x^2 + \alpha_y^2 + \alpha_z^2 - 3\right], \tag{3}$$

where B is temperature and gel crosslinking concentration dependent. The classical way of measuring gel shear modulus involves anisotropical deformation of the sample (e.g., uniaxial elongation). Assuming the gel is constrained (or controlled) in d dimensions and can swell isotropically in other $3\text{-}d$ directions, then using Eq. (2), F_2 can be written as

$$F_2 = B \left[d\alpha^2 + (3 - d) \left(\frac{\phi}{\phi_0} \right)^{-2/(3-d)} \alpha^{-2d/(3-d)} \right] \tag{4}$$

Where α is the deformation ratio in the constrained directions. The unimportant constant in F_2 has been omitted in the above equation for simplicity.

The swelling osmotic pressure of the gel under constraint α=constant is

$$\begin{aligned} \omega &= - \left(\frac{\partial F}{\partial V} \right)_\alpha \\ &= \Pi - \frac{2B}{V_0} \left(\frac{\phi}{\phi_0} \right)^{(1-d)/(3-d)} \alpha^{-2d/(3-d)} \end{aligned} \tag{5}$$

where $\Pi(\phi)$ is the osmotic pressure contribution from F_1.

In order to study the concentration dependence of the mechanical properties, let us define Λ the constraint relative to the isotropic state with network concentration equal to ϕ:

$$\Lambda = \alpha \left(\frac{V_0}{V} \right)^{1/3} = \alpha \left(\frac{\phi}{\phi_0} \right)^{1/3}, \tag{6}$$

then

$$F_2 = B \left(\frac{\phi}{\phi_0} \right)^{-2/3} \left[d\Lambda^2 + (3 - d)\Lambda^{-2d/(3-d)} \right] \tag{7}$$

In a pure shear deformation, the volume, and therefore ϕ, are kept constant. The pressure that causes the deformation is then equal to,

$$\begin{aligned} P &= -\frac{1}{dV} \left(\frac{\partial F_2}{\partial \Lambda} \right)_\phi \\ &= -\frac{2B}{V_0} \left(\frac{\phi}{\phi_0} \right)^{1/3} \left[\Lambda - \Lambda^{-(3+d)/(3-d)} \right]. \end{aligned} \tag{8}$$

The factor in front of the bracket defines the shear modulus G,

$$G = G_e \left(\frac{\phi}{\phi_e} \right)^m \tag{9}$$

where G_e is the shear modulus of the unconstrained gel at equilibrium,

$$G_e = \frac{2B}{V_0} \left(\frac{\phi_e}{\phi_0} \right)^m \tag{10}$$

The exponent m characterizes the concentration dependence of the shear modulus for a given gel. In Flory-Huggins theory, $m=1/3$.

Using Eqs. (6), (9), and (10), Eq. (5) becomes

$$\omega = \Pi - G_e \left(\frac{\phi}{\phi_0} \right)^{1/3} \Lambda^{-2d/(3-d)} \tag{11}$$

Taking the scaling approach [11, 12, 13], the osmotic contribution from the mixing term (F_1) can be written as

$$\Pi = B' \left(\frac{\phi}{\phi_e} \right)^n \tag{12}$$

where B' is concentration independent. When the gel is unconstrained and at swelling equilibrium ($\phi=\phi_e$ and $\Lambda=1$), the osmotic pressure ω is zero. This requirement yields $B' = G_e$, therefore,

$$\omega = G_e \left[\left(\frac{\phi}{\phi_e} \right)^n - \left(\frac{\phi}{\phi_e} \right)^m \Lambda^{-2d/(3-d)} \right] \tag{13}$$

$$P = G_e \left(\frac{\phi}{\phi_e} \right)^m \left[\Lambda - \Lambda^{-(3+d)/(3-d)} \right] \tag{14}$$

For constrained gel ($\Lambda \neq 1$) at equilibrium, substitute $\omega=0$ in Eq. (13) and we have

$$\Lambda = \left(\frac{\phi}{\phi_e} \right)^{-(n-m)(3-d)/2d} \tag{15}$$

Introducing a new exponent δ,

$$\delta = n - m \tag{16}$$

then from Eq. (15),

$$\delta = -\frac{2d}{(3-d)} \frac{\ln(\Lambda)}{\ln(\phi/\phi_e)} \tag{17}$$

This relation can be used to obtain the value of δ once the volume reduction (or, ϕ/ϕ_0) and deformation Λ are known.

It is convenient to introduce the bulk modulus K of the gel,

$$\begin{aligned} K &= \left. \phi \frac{\partial \omega}{\partial \phi} \right)_{\Lambda=1} \\ &= G_e \left[n \left(\frac{\phi}{\phi_0} \right)^n - m \left(\frac{\phi}{\phi_0} \right)^m \right] \end{aligned} \tag{18}$$

Using this equation, one find $K_e = G_e(n-m)$ at equilibrium ($\phi=\phi_e$ and $\Lambda=1$), or

$$\delta = \frac{K_e}{G_e} \tag{19}$$

Therefore, the physical meaning of the exponent δ is that the value of δ is equal to the ratio of gel network moduli. The table below summarizes the results for different d.

	ω	P	exponent δ
free swelling: $d=0$	$G\left[(\phi/\phi_e)^\delta - 1\right]$	0	–
1-d compression	$G\left[(\phi/\phi_e)^\delta - \Lambda^{-1}\right]$	$G\left[\Lambda - \Lambda^{-2}\right]$	$\ln(\Lambda^{-1})/\ln(\phi/\phi_e)$
2-d compression	$G\left[(\phi/\phi_e)^\delta - \Lambda^{-4}\right]$	$G\left[\Lambda - \Lambda^{-5}\right]$	$\ln(\Lambda^{-4})/\ln(\phi/\phi_e)$

Non-ionic gels in good solvent can be considered as semidilute systems [13]. For these systems, it has been shown theoretically that $n = 9/4$ and $m = 1/3$ [12, 13]. These results have been intensively verified experimentally for both organogels [14, 15] and hydrogels [16, 17]. In particular, Geissler, et. al. showed that for polyacrylamide gels in water, $n = 1.9$ and $m = 0.34$ [17].

Although there is a lack of direct study of the mechanical properties of polyelectrolyte gels, the osmotic pressure of polyelectrolyte solutions is well understood. There are several length scales that are important to the behavior of a polyelectrolyte solution. They are the ionic group spacing A of the polymers, Debye-Huckle screening length κ^{-1}, and Bjerrum length Q. Odijk [18] proposed to consider polyelectrolytes as worm-like chains with total persistence length L_t, which equals to the summation of polymer intrinsic persistence length L_p and the electrostatic contribution L_e, i.e.,

$$L_t = L_p + L_e = L_p + \frac{Q}{4\kappa^2 A^2 f^2} \tag{20}$$

where f is ion condensation factor. The osmotic pressure of the solution is

$$\Pi \sim (L_t/\kappa)^{3/4}(AC)^{9/4}, \tag{21}$$

where C is the monomolar concentration of the polymer in solution. In pure water (without added salt), $\kappa^2 = 4\pi AC$ and $L_t \sim L_e$,

$$\Pi \sim C^{9/8}. \tag{22}$$

With excess salt, $\kappa^2 (= 8\pi QA)$ is independent of C and

$$\Pi \sim C^{9/4} \tag{23}$$

Therefore the exponent n of polyelectrolyte gels in water with and without salt is expected to be equal to 9/4 and 9/8, respectively. Recently Wang and Bloomfield [19] have successfully applied the renormalization group theory to study semidilute polyelectrolyte solutions, such as poly(styrene sulfonate). They found that the existing experimental results agree well with theoretical predictions given by equations (22) and (23) over a wide range of salt concentration, polymer molecular weight, and polymer concentration.

Given these polyelectrolyte solution results, it is interesting to see how the corresponding polyelectrolyte gels behave. Assuming the exponent m of polyelectrolyte gel is $1/3$, then from Equations (13), (22), and (23) we can obtain the theoretically predicted exponent δ. In the table below, we tabulated these results for polyelectrolyte gels together with non-polyelectrolyte gels for comparison.

	Non-polyelectrolyte Gel			Polyelectrolyte Gel		
	n	m	δ	n	m	δ
Without salt	9/4	1/3	1.92	9/8	1/3	0.8
With excess salt	9/4	1/3	1.92	9/4	1/3	1.92

As shown in this table, for polyelectrolyte gel in pure water, $\delta = 0.8$. Once excess amount of salt is added, the δ value is expected to be approaching 1.92.

Geissler et al [20] recently demonstrated the effect of gel inhomogeneity on the swelling osmotic pressure. This effect was neglected in this chapter. However, it can be incorporated in the theory outlined earlier easily without changing the general conclusions of this paper. We also have neglected the non-Gaussian elasticity treatment [4] for the sake of simplicity. The finite chain extension (Langevin model) effect will result in a non-scaling elastic free energy, which has a higher ($\dot{\iota}1/3$) "apparent exponent" m for any given small region of ϕ. Since all of our numerical results reported here were obtained from samples with small swelling ratios ($V_e/V_1\dot{\iota}4$, or linear expansion less than 60%), we expect the Flory-Huggins elasticity to be applicable.

Experimental

The details of the experimental method can be found in previous publications [9]. The principle of the experimental technique is based on Eq. (17) with $d=2$. The two dimensional constraint is provided by making gel films on top of non-swellable substrates. These films can swell only in the direction perpendicular to the film surface. The experiment involves the measurement of the thicknesses of constrained and free gel films. From these measurements the polymer volume fraction reduction ϕ/ϕ_e and shear deformation Λ can be calculated. The exponent δ can then be obtained by using Eq. (17).

The degree of ionization of the samples is quantified in terms of the ratio of monomer number concentrations of sodium acrylate to that of acrylamide. The unit used is mM (for milli-mole per mole).

Results and Discussions

In all of our ionic gel samples, the degree of ionizations are all well below the point at which the ion condensation occurs. Therefore we will not discuss its potential impact here.

Non-ionic Gels. Figure 1 shows the exponent δ of polyacrylamide gels versus their equilibrium network concentration ϕ_e. The ϕ_e was changed by varying the chemical composition of the pre-gel solutions. Two approaches were taken: one was to change the amount of crosslinking, the other was to change the amount

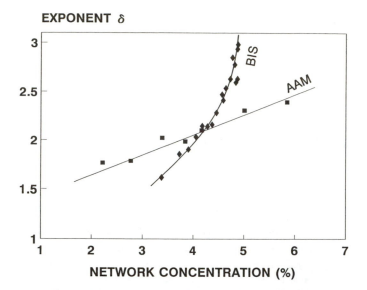

Figure 1:   The scaling exponent δ of polyacrylamide gels as a function of
network concentration ϕ_e. The two series of data were obtained from samples with varying crosslinking concentration (BIS) and acrylamide monomer
concentrations (AAM) in the pre-gel solution, respectively. Notice these are
non-ionic gels.

of acrylamide monomer concentration. As shown in Figure 1, the exponent δ is
chemical composition dependent [21]. The value of δ varied from 1.75 to 2.4 for
the series with acrylamide concentration varied, and from 1.65 to 3 for the series
with crosslinkers varied. In both cases, the higher the network concentration ϕ_e,
the higher the δ. At high network concentration, the system deviates from the
semidilute model, resulting in higher δ value than what the theory expects. This
deviation is increased further by the hydrophobicity of the crosslinkers used (N,N'-
methylenebisacrylamide) for the highly crosslinked samples. From this figure, the
samples with ϕ_e in the range of 3∼4% can be considered ideal semidilute systems
in good solvent.

    Combining the compression and osmotic deswelling methods, Geissler, et al [17]
measured both n and m of acrylamide gels. The sensitivity of δ to the change
of chemical composition was not observed in their results. Detailed discussion
regarding the non-ionic gel results can be found in ref [21].

Effect of Degree of Ionization.   As sodium acrylate was introduced into the
network, the gel became ionic and the equilibrium degree of swelling increased
(ϕ_e decreased). This is shown in Figure 2a. At the same time, the behavior
of δ changed markedly, as shown in Figure 2b. The δ value started at 2.1 and
dropped down to 1.08 when 2 mol% of the NaAc was used (the swelling ratio
at this point is: $V_e/V_1=4$). Further increase of sodium acrylate created surface
patterns on the constrained gels and therefore the δ values calculated may not

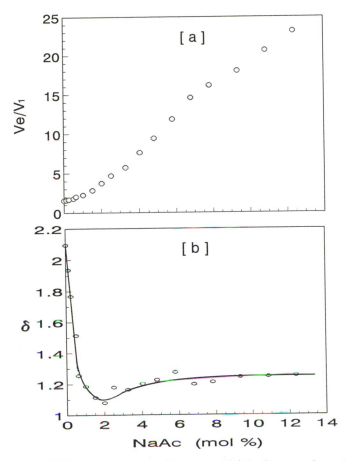

Figure 2: [a] The equilibrium swelling ratio V_e/V_0 of ionic polyacrylamide gels in water as a function of degree of ionization. [b] The value of δ as a function of degree of neutralization in water. The δ values decreases quickly as ionic groups are introduced into the network.

be as reliable. Based on the steep slope, it is conceivable that further increase of degree of ionization could produce even lower δ value. This indicates that the δ of the ionized polyacrylamide is at most equal to 1.08 for NaAc concentration higher than 2 mol%, or the exponent n is at most equal to 1.4 (assuming $m=1/3$). This number apparently is approaching the theoretical result of $n=1.12$ [19]. It is worth pointing out that it takes only a very low concentration of ionic groups on the network to make the network behave like a polyelectrolyte one.

From figure 2a, the network concentration of the 2mol% ionized gel was around 0.4 monomolar (3% by weight). The concentration of the ionic group is therefore 8×10^{-3} molar. Using the ideal gas approximation, then the osmotic pressure of from the counterions (Na^+) is,

$$\Pi \approx 0.18 \ atm = 1.8 \times 10^5 \ dynes/cm^2 \qquad (24)$$

Although the ideal gas approximation exaggerates the electrostatic interaction, the above value is far greater than the osmotic pressure of the corresponding polyacrylamide polymer solution, which is estimated to be around $2.4 \times 10^4 dynes/cm^2$ [17]. This explains the reason why a small amount of ionic groups on the network can make the network a totally different system.

Effect of Salt Concentration.   As sodium chloride salt is added to the solvent, the screening effect reduces the electrostatic interaction dramatically. As has been argued by Odijk [18], with sufficiently high salt concentration, the osmotic pressure of polyelectrolyte solutions should behave like non-polyelectrolyte [19]. Figure 3 compares the δ as a function of NaAc concentration with and without salt added. With 10 mMol salt added to the solvent, the δ value was raised to 1.8.

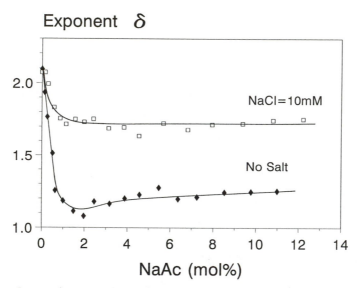

Figure 3:   A comparison of the value of δ in water and aqueous (10 millimolar) salt solution. With salt, the δ values increase from the no salt situation to a much higher level.

Figure 4 is the results of several gels with fixed amount of NaAc in salt solution. At low salt concentration (Figure 4a), the δ value increases first and then reaches different plateau levels depending on the degree of gel ionization. As the concentration of NaCl is further increased (Figure 4b), the δ value appears to be converging asymptotically to 1.8.

Four Regions.   Combine the effects of both ionization and salt concentration, the behavior of δ value can be summarized as shown in Figure 5. Four distinct regions have been identified for the poly(acrylamide/sodium acrylate) gel system. These regions correspond to different phenomenon and may require different theories to explain. The first region is the transition region from no-salt to with-salt. Its boundary line can be defined as the salt and ionization concentrations at which the δ value reaches certain level (e.g., 80% of its full range). The

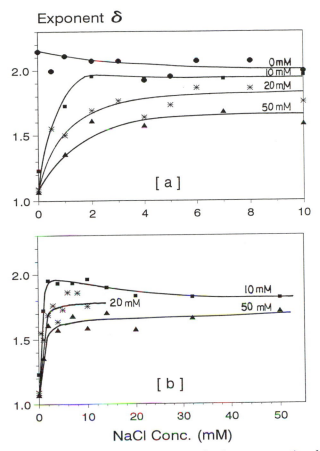

Figure 4: The exponent δ as a function of salt concentration for gels of different degree of ionization (0mM, 10mM, 20mM, and 50mM). [a] In this salt concentration region, all samples reach a plateau. [b] As the salt concentration is further increased, the δ values of different ionic gels converge asymptotically to the same value ($\sim$1.8).

value of δ changes the most in this region. The second region is the region in which the δ exhibits a peak. In the third region, the δ value is slightly dependent on ionic strength and salt concentration. Once reaching the fourth region, the δ value of the polyelectrolyte gel system stays at about 1.8. Figure 5 also shows a maximum-δ line corresponds to the maxima observed in Figure 4b. This line terminates at ionization between 20mM and 50mM.

Conclusions

In conclusion, the ratio of bulk modulus to shear modulus, K/G, of ionic polyacrylamide (PAAM) gels is studied as functions of degree of ionization and salt concentration. This ratio is directly related to exponent $\delta = (n - m)$, with n and m the exponents characterizing the concentration dependence of bulk and

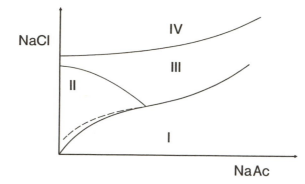

Figure 5:   The four regions of ionic polyacrylamide gels in the presence of NaCl.

shear modulus, respectively. The value of δ depends on the network ionization and solvent salt concentration. In pure water (without added NaCl), $\delta < 1.08$ for $\alpha > 0.02$. When excess amount of NaCl is added, the δ value recovers to about 1.8. The experimental results are in good agreement with theoretical predictions.

It was also observed that the ionization-salt concentration space can be divided into four regions. The region-I and III are transitional regions. In region-II, a peak in δ exists. The last region corresponds to polyelectrolyte gels in excess salt. In this region, the δ value is about 1.8, in agreement with the theoretically predicted value of 1.9.

Acknowledgement is made to the Donors of The Petroleum Research Fund, administered by the American Chemical Society, and to the U.S. Army Research office under Grant No. DAAH04-93-G-0215 for support of this research.

References

[1] Brannon-Peppas, L.; Peppas, N. A. *Int. J. Pharm.* **1991**, *70*, 53.

[2] Tanaka, T.; Nishio, I.; Sun, S-T.; Ueno-Hisho, S. *Science* **1982**, *218*, 467.

[3] Osada, Y.; Ross-Murphy, S. B. *Scientific American*, **1993**, *May*, 82.

[4] R. K. Prud'homme and Y. Yin, *ACS Polym. Mat. Sci. Eng.* **1993**, *69*, 527.

[5] Shiga, T.; Hirose, Y.; Okada, A.; Kurauchi, T. *J. Appl. Polym. Sci.* **1992**, *44*, 249.

[6] Tanaka, T.; Sun, S-T.; Hirokawa, Y.; Katayama, S.; Kucera, J.; Hirose, Y.; Amiya, T. *Nature* **1987**, *325*, 796.

[7] Li, Y.; Tanaka, T. *J. Chem. Phys.* **1989**, *90*, 5161.

[8] Hirotsu, S. *Macromolecules* **1990**, *23*, 905.

[9] Li, Y.; Li, C.; Hu, Z. *J. Appl. Polym. Sci.* **1993**, *50*, 1107.

[10] Flory, P. J. *The Principles of Polymer Chemistry*, Cornell University Press, Ithaca and London, 1953.

[11] De Gennes, P. G. *Scaling Concepts in Polymer Physics*, Cornell University Press., 1979.

[12] Schaefer, D. W. *Polymer* **1984**, *25*, 387.

[13] Candau, S.; Bastide, J.; Delsanti, M. *Adv. Polym. Sci.* **1982**, *44*, 27.

[14] Horkay, F.; Zrinyi, M. *Macromolecules* **1982**, *15*, 1306; **1988**, *21*, 3260.

[15] Munch, J. P.; Candau, S.; Herz, J.; Hild, G. *J. Phys. (Paris)* **1977**, *38*, 971; Candau, S. J.; Young, C. Y.; Tanaka, T.; Lemarechal, P.; Bastide, J. *J. Chem. Phys.* **1979**, *70*, 4694.

[16] Geissler, E.; Horkay, F.; Hecht, A-M. *Macromolecules* **1991**, *24*, 6006.

[17] Geissler, E.; Hecht, A-M.; Horkay, F.; Zrinyi, M. *Macromolecules* **1988**, *21*, 2594.

[18] Odijk, T. *J. Polym. Sci., Polym. Phys. Ed.* **1977**, *15*, 477; Odijk, T.; Houwaart, A. C. *J. Polym. Sci., Polym. Phys. Ed.* **1978**, *16*, 627; Odijk, T. *Macromolecules* **1979**, *12*, 688.

[19] Wang, L.; Bloomfield, V. *Macromolecules* **1990**, *23*, 804; and **1990**, *23*, 194.

[20] Geissler, E.; Horkay, F.; Hecht, A-M.; Zrinyi, M. *J. Chem. Phys.* **1989**, *90*, 1924; Horkay, F.; Hecht, A-M.; Mallam, S.; Geissler, E.; Rennie, A. R. *Macromolecules* **1991**, *24*, 2896; Hecht, A-M.; Guillermo, A.; Horkay, F.; Mallam, S.; Legrand, J. F.; Geissler, E. *Macromolecules* **1992**, *25*, 3677.

[21] Hu, Z.; Li, C.; Li, Y. *J. Chem. Phys.* **1993**, *99*, 7108.

RECEIVED April 13, 1994

Chapter 6

Rate-Limiting Steps for Solvent Sorption and Desorption by Microporous Stimuli-Sensitive Absorbent Gels

Bhagwati G. Kabra[1] and Stevin H. Gehrke

Department of Chemical Engineering, University of Cincinnati, Cincinnati, OH 45221–0171

Microporous responsive gels can swell or shrink in response to an environmental stimulus orders of magnitude faster than comparable non-porous gels. In non-porous gels, the rate-limiting step for sorption or desorption is the diffusion of the network through the solvent from one equilibrium conformation to another. However, the characteristic length for diffusion reduces to a micron scale in a microporous gel, thus reducing the network response time drastically. Since convective flow into or out of the gel through the micropores can also be very fast, the upper bound on the response rate (and the minimum response time) is set by the rate of stimulus transfer. These concepts were demonstrated by examining the mass changes of microporous and non-porous hydroxypropyl cellulose gels in response to changes in temperature, salt concentration and solvent composition.

Crosslinked polymer gels that undergo large changes in swollen volume with changes in their solution environment have been termed "stimulus-sensitive" or "responsive" gels (1). Stimulus-sensitive gels have been used as recyclable, selective absorbents to dewater coal slurries, biochemical solutions and waste sludges. Their potential as switches, mechano-chemical actuators, or components of drug delivery systems has also been demonstrated (1,2). The dewatering application for responsive gels operates in the following manner. The gel is added to the solution or slurry in its shrunken state. The stimulus is applied (for example, by changing the solution temperature or pH), causing the gel to swell and absorb solvent and low molecular weight species, while excluding large molecules and particles. Removal of the gel leaves behind a concentrate; the gel is regenerated by reversing the stimulus, causing the gel to shrink and expel the absorbed solution (1,3-6). Recently a company has been founded to manufacture and market these materials (7).

An important parameter in the development of applications using these adjustable absorbency gels is the response kinetics. For example, the amount of gel needed to dewater a given amount of slurry is directly proportional to the cycling time of the gel.

[1]Current address: Alcon Laboratories, Inc., 6201 South Freeway, Fort Worth, TX 76134–2099

0097–6156/94/0573–0076$08.00/0
© 1994 American Chemical Society

This paper examines the ranges of response rates that can be obtained for the most-studied responsive gel stimuli: temperature, ionic strength, solvent quality and pH. The goal is to identify the rate-limiting steps for the swelling and shrinking processes of responsive gels generally and to determine the upper and lower bounds that these steps place on the response rates.

Concepts

The first step in the volume change of wetted stimulus-sensitive gels in response to an environmental change is the transfer of the environmental stimulus throughout the gel. This change will cause the gel to swell or shrink by some combination of diffusion and convection of the solvent through the gel. In temperature-sensitive gels, heat conduction is the usual mechanism of stimulus transfer. The thermal diffusivity (α) of a hydrated gel is on the order of 10^{-3} cm^2/s *(8-10)*. For salt- or solvent-sensitive gel, the diffusion coefficient of the salt or solvent in the gel will be the key parameter. These coefficients are on the order of 10^{-5} cm^2/s. The rate of sorption or desorption of the solvent in response to these stimuli may also be limited by the rate of network movement through the solvent. It is generally accepted that the polymer chains in a crosslinked gel move through a solvent by a cooperative diffusive process *(11-12)*. Thus the movement of polymer chains in responsive gels, induced by a stimulus, occurs with mutual or cooperative diffusion coefficients on the order of 10^{-7} cm^2/s. Convection can be very fast and can even alter the rate of stimulus transfer. In gels, convective flow will enhance the stimulus rate if solvent and stimulus are moving in the same direction; conversely, the stimulus rate will be slowed if the convective flow opposes the stimulus transfer. The rate of convection will depend upon the gel microstructure, especially pore size, porosity and pore interconnectivity.

However, if the gel network is uniform down to a submicron scale, significant convection is unlikely to occur. In the absence of convection, the characteristic dimension L for diffusive processes will be that of the entire sample: radius for a sphere or long cylinder and half-thickness for a flat sheet. The order of magnitude of time required for a diffusion-controlled process to occur can be estimated as $t = L^2/D$ *(13)*. Thus the time required for a network diffusion-controlled volume change of a responsive gel with L = 1 mm can be estimated as $(0.1 \text{ cm})^2/(10^{-7} \text{ cm}^2/\text{s}) = 10^5$ s, or about a day. In contrast, the time required for equilibration with a solvent or salt change is only 10^3 s, or about 15 minutes. By the same logic, the time required to reach thermal equilibrium is only 10 s. Thus if the characteristic dimensions for all of the potential rate-limiting steps are the same, diffusion of the network is expected to be the rate-limiting step for all types of responsive gels.

From this discussion it is apparent that the rate of network motion is usually the rate limiting step in non-porous gels, as it occurs 2 to 4 orders of magnitude more slowly than the rate of stimulus change. Some exceptions to this general observation exist; for example, external mass transfer of ions can be rate-limiting for pH-sensitive gels in dilute, unstirred solutions *(14,15)*. But in fact, for homogeneous, non-porous gels, this slow, network diffusion-controlled response is typically observed.

To increase the response rate, we hypothesized that microporous gels with open-celled microstructures would have much faster response rates than non-porous gels *(1,8,16-18)*. This would be true if the characteristic dimension for network diffusion is reduced to microns, reducing the network response time to fractions of a second, while interconnected pores would allow convective flow of the solvent into or out of the gel. Under these conditions, the rate of stimulus change is likely to become the rate limiting step because the characteristic dimension for stimulus transport would remain the macroscopic dimension (since convection follows the stimulus). Convection would be rate-limiting only within a window of pore size and porosity that left it slower than the stimulus transfer rate yet faster than network diffusion. However, gels with very small pores or low porosities would remain network diffusion-limited.

To test these ideas, we developed a technique that could produce such stimuli-sensitive gels with a wide variety of microstructures (8,18). This process involves chemical crosslinking of a polymer as it undergoes temperature-induced phase-separation. With these microporous gels, we could answer the question "in principle, what is the maximum response rate that could be achieved with responsive gels, irrespective of the technique used to produce the microstructure?" Thus in this paper, we wish to verify the validity of the concepts presented in this section and to establish the lower and upper bounds on response rates for gels responsive to temperature, ionic strength and solvent change.

Experimental

The gel model used in this paper is synthesized from hydroxypropyl cellulose (HPC) crosslinked with divinyl sulfone in alkaline solution as the polymer phase-separates upon heating above its lower critical solution temperature (LCST). The microstructure of the resulting gel depends upon the precise conditions of synthesis and varies from non-porous (homogeneous) to porous at micron length scales. The samples tested were made as flat sheets about 1 mm thick and with aspect ratios greater than 10. Water-swollen HPC gels shrink as either temperature, ammonium sulfate concentration or acetone concentration are increased (8,18,19). The effects are reversible. The concepts being developed here are independent of the specific gel used and thus readers interested in the details of synthesis of these gels are referred to other sources (1,8,16-19).

The specific sample (I.D. #P46) chosen for the transport studies in this paper was formed from a 9 wt.% solution of HPC in pH 12.2 NaOH including 1.6 wt.% of the crosslinker, divinyl sulfone. The solution was poured into a mold made by clamping glass plates together that were separated by silicon rubber gaskets. After 8.9 minutes of reaction at room temperature, the mold was immersed into a water bath at 46.5 °C, above the LCST of HPC, which is 42 °C. After 1.7 minutes of reaction in the phase-separated state, the mold was removed from the water bath and the reaction was allowed to proceed for an additional 2390 minutes at room temperature. The mold was then opened and an opaque gel sheet was recovered. The sol fraction was leached from the gel by soaking in alternate hot and cold water baths over a period of several days.

The response rates were measured using either videomicroscopy or a gravimetric technique. These techniques are described in detail elsewhere (8,16-17). In order to determine the rate governing steps, the rates of different steps were determined or estimated independently. The rate of network motion could be estimated using a diffusion analysis using either a conventional Fick's law approach or using a mathematically similar theory based on the equations of motion (1,8,9,14,17,20,21). For thermal response, the rate of temperature change can be estimated via heat conduction theory using the measured value of thermal conductivity (8-10,22). For salt and solvent response, the rate of concentration change in the gel can be estimated assuming Fickian diffusion using known values of the diffusion coefficients of salt and solvent in water, reduced by a tortuosity factor (8,14,22).

The rate of convection in the absence of rate-limiting stimuli was determined as follows. A microporous HPC gel was shrunken to equilibrium at 60 °C, removed from solution, blotted free of surface water, and allowed to cool to room temperature in air. This eliminates the rate of stimulus change as a factor. Thus when the gel is immersed in water at room temperature, its rate of sorption will be the rate of convection, assuming that the rate of microstructure sorption is much faster than the macroscopic sorption rate. This will be known to be the case if the sorption is very rapid, since convection typically occurs orders of magnitude faster than diffusion.

Results and Discussion

Scanning electron micrographs of freeze-dried samples of the HPC gels formed by crosslinking the polymer solution as it underwent thermally-induced phase separation showed that microporous gels could be produced under appropriate synthesis conditions *(8,18)*. The micrograph of gel sample P46 shown in Figure 1 demonstrates that the gel was microporous, with interconnected pores in the range of 0.5 to 8 μm in diameter and strut thicknesses ranging from 0.5 to 1.0 μm. The porosity of this gel is 0.64, as estimated based on the amount of water that could be squeezed from the gel under a briefly applied pressure of approximately 15 psi. The swelling and shrinking rates of this gel were measured in response to temperature change, ammonium sulfate concentration change and solvent quality change (increasing acetone concentration makes the solvent poorer for HPC). For contrast, the swelling and shrinking kinetics of a non-porous gel (sample ID #P73) made from the same starting solution composition were also measured *(8)*.

These data are given in Figures 2 - 5, in which the magnitude of water uptake or loss is plotted against time in a normalized fashion as M_t/M_∞, where M_t is the mass of solvent absorbed or desorbed at time t and M_∞ is the total mass absorbed or desorbed as t→+∞. These results are used to identify the upper and lower bounds which can be observed for responsive gels generally.

Response to Changes in Temperature. Figures 2 and 3 demonstrates that microporosity dramatically increases the rate of swelling and shrinking in response to a temperature change, as hypothesized. Figure 2 shows that the microporous HPC gel sample reaches equilibrium swelling or shrinking within 15 seconds, although the sample is over 1 mm thick. This rate of equilibration is a thousand times faster than required for the non-porous HPC gel shown in Figure 3.

To better understand the mechanism by which microporosity increased the rate of sorption, additional data analysis was undertaken. The data in Figure 3 were compared against the diffusion curves that best fit the data. This can be done by the procedure described in several references from our group *(8,17,23)*. The coefficients extracted are on the order of 10^{-7} cm^2/s, values characteristic of volume change controlled by the rate of network motion *(1,8,12,17,20,24)*. This represents a lower bound on the rate of sorption or desorption for responsive gels.

For the non-porous gel, the characteristic dimension for diffusion is the half-thickness of the sheet, but for the microporous gel, the characteristic dimension should be closer to that of the pore walls or struts. Using the average strut thickness as the characteristic dimension and a diffusion coefficient of 10^{-7} cm^2/s, the time required for network motion in the microporous gel sample P46 is estimated to be less than 0.1 s. Since this is much less than the time required for even this gel to swell or shrink, the rate-limiting step appeared to be either the rate of heat transfer or the rate of convection. The rate of convection was experimentally measured for swelling as described in the experimental section and the rate of heat transfer was estimated using heat conduction theory and the measured value of gel thermal conductivity. These were then plotted in Figure 2 for comparison to the swelling and shrinking curves. Convection occurred so quickly - within a couple seconds- that most of the data couldn't be collected. However, the rate of heat transfer is slower and appears to be the rate-limiting step upon swelling.

Since the rate of heat transfer is by far the fastest of the stimuli tested, and since the rate of heat conduction will be virtually independent of microstructure (since the thermal diffusivities of the gel and water are similar), it places the upper bound on the

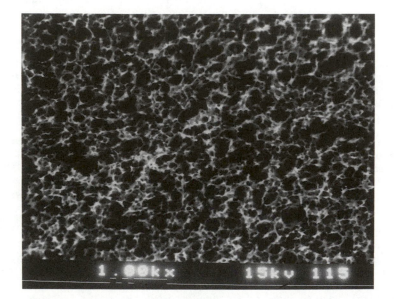

Figure 1: Scanning electron micrograph of microporous gel sample P46, 1000×
magnification. The pores are interconnected and range in size from 1 to 8 μm;
strut thicknesses range from 0.5 - 1.0 μm.

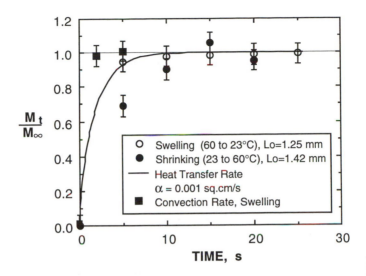

Figure 2: Normalized mass (measured) and centerline temperature (calculated) changes of microporous HPC gel sample P46, induced by the indicated changes in temperature in the surrounding water. The rate of heat transfer dominates the volume change kinetics of temperature sensitive microporous HPC gels; convection is faster.

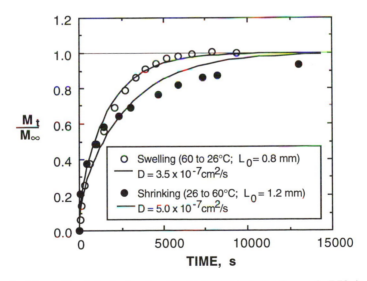

Figure 3: Normalized mass changes of non-porous HPC gel sample P73, induced by the indicated changes in temperature in the surrounding water. The volume change kinetics of this gel are governed by the rate of network motion. Swelling and shrinking can be approximately described by diffusion analyses.

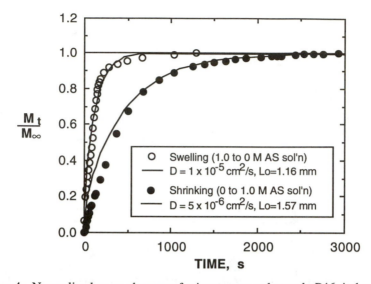

Figure 4: Normalized mass changes of microporous gel sample P46, induced by the indicated changes in ammonium sulfate (AS) concentration in the surrounding solution. The volume change kinetics are dominated by the rate of change of salt concentration inside the gel.

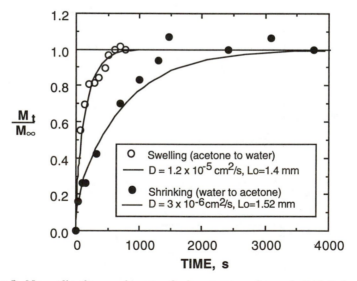

Figure 5: Normalized mass changes of microporous gel sample P46, induced by the indicated changes in acetone concentration in the surrounding solution. The volume change kinetics are dominated by the rate of change of solvent concentration inside the gel.

response rate of these gels. In fact, numerous other temperature-sensitive HPC gels with different microstructures were tested; the rates varied over a continuum between the upper and lower bounds identified in examples in Figs. 3 and 4. However, due to the solvent convection, convective heat transfer will occur and thus there may be some dependence of heat transfer on the microstructure *(8)*. Strong evidence of the coupling of convection and stimulus transfer is seen with the salt and solvent-sensitive gels discussed below.

Response to Changes in Salt Concentration. Water-swollen HPC gel at room temperature shrinks when immersed in a solution of ammonium sulfate due to the disruption of hydrogen bonding between the polymer and solvent and enhancement of hydrophobic interactions between polymer chains *(25)*. The volume change kinetics of gel P46 in response to a change in the concentration of ammonium sulfate solution between 0 and 1.0 M is given in Figure 3. As seen by comparing Figures 3 and 4, this response is about 50 to 100 times faster than what is seen for the non-porous gel in response to temperature. The shrinking and swelling kinetics approximately fit diffusion curves with diffusion coefficients of magnitudes 5×10^{-6} cm^2/s and 1×10^{-5} cm^2/s, respectively. Thus the rates lie between the rate of network diffusion and the rate of convection as identified for temperature-sensitive gels. Convection and network diffusion should occur at about the same rates in this experiment as in the temperature experiment. Thus rate of salt transport appears to be the rate-limiting step.

The heterogeneity of the microporous gels makes analysis of mass transfer within them complex. But for the purposes of identifying the rate-limiting steps only, the following estimation is sufficient. First of all, the diffusion coefficient of ammonium sulfate is estimated from the ionic diffusion coefficients as described by Cussler, yielding a value of 1.5×10^{-5} cm^2/s *(13)*. The diffusion coefficient of solutes within the gel can be estimated using the following relationship *(26)*:

$$D = D_o \left(\frac{1 - \phi_p}{1 + \phi_p} \right)^2 \tag{1}$$

where: D = diffusion coefficient in the gel.
D_o = diffusion coefficient in the solution.
ϕ_p = polymer volume fraction.

The polymer volume fractions in the shrunken (1 M ammonium sulfate) and swollen (distilled water) states were 0.26 and 0.11, respectively. Using equation 1, the average diffusion coefficient of ammonium sulfate in the gel was thus estimated to be 7×10^{-6} cm^2/s. This value is intermediate between the diffusion coefficients obtained by fitting the swelling and shrinking to a diffusion equation, supporting the idea of salt transport as the rate-limiting step.

Evidence of influence of convection on the rate of salt transport is seen in the fact that the swelling occurs more quickly than shrinking for microporous gels. This is the opposite of what is observed for non-porous gels responding to temperature, salt or solvent stimuli (after accounting for effects due to differences in thickness and time lags caused by skin formation) *(27)*. In this case, swelling occurs about twice as fast as expected if salt diffusion was rate-limiting, while shrinking occurs about half as fast as expected.

The influence of convection on mass transfer can explain this observation, as follows. During the shrinking process, volume change is caused by diffusion of ammonium sulfate ions in the gel. As the ammonium sulfate concentration increases at a given point in the gel, that part of the gel shrinks, causing a convective flow of solution from the gel to the outside solution. This convective flow of solution is in the direction opposite to the stimulus. As a result, the convection of solution retards the rate of stimulus change. In contrast, during the swelling process, ion-free water is absorbed into the gel, decreasing the ion concentration inside the gel. This results in gel swelling and pulls more ion-free water into the gel by convective flow. Thus, during swelling, convection moves in the same direction as the stimulus and hence enhances the rate of stimulus change (although the diluted ions must diffuse outward eventually in order to achieve equilibrium with the water; this might explain the lag seen in Figure 3 for the last 10% of the swelling curve).

Therefore, the rate of ion transport places the upper bound on the rate of volume change for salt-sensitive responsive gels. This bound is approximately equal to the rate of diffusion of the ions into the gel, with the sample dimension as the characteristic length. However, this upper bound is not independent of gel microstructure as the convection does influence the rate of stimulus change. The fastest swelling rates will be observed for highly microporous gels with large pores, whereas the fastest shrinking rates will be observed for gels with microstructures in which the rate of convection is not significantly faster than the rate of diffusion of ions.

Although not explicitly demonstrated, the lower bound on the response rate should be the diffusion rate of network, as seen for temperature-sensitive gels. Akhtar has shown that for non-porous poly(N-isopropylacrylamide) gels, network diffusion is invariably rate-limiting and the diffusion coefficient is not a significant function of the stimulus *(27)*. Theoretically, changes in solution could affect the network diffusion coefficient via changes in polymer concentration, polymer-solvent friction (viscosity), and chain conformation. Such effects should not affect the order of magnitude of the diffusion coefficient, except in unusual cases; perhaps crossing a discontinuous phase transition, for example. Thus the lower bound expected for salt-sensitive gels should be described by a diffusion coefficient of the order of 10^{-7} cm^2/s (about 50-100 times smaller than the upper bound).

It should be noted that this analysis applies strictly to a nonionic salt-sensitive gel. Ionizable salt-sensitive gels may be subject to other rate-limiting mass transfer steps such as those described under the pH-sensitive section below.

Response to Changes in Solvent Composition. HPC gels swell less in acetone than in water due to less favorable interactions between polymer and solvent and thus change their volumes in response to changes in solvent between water and acetone. The volume change kinetics of gel P46 in response to these changes are shown in Figure 5. The shrinking and swelling kinetics can be approximately fitted to diffusion curves with diffusion coefficients of magnitude 3×10^{-6} cm^2/s and 1.2×10^{-5} cm^2/s, respectively.

The analysis of this process proceeds in the same manner as for the salt-sensitive case. The polymer volume fractions in the swollen and shrunken stages were 0.11 an 0.14, respectively. Thus binary diffusion coefficient of acetone and water within the gel was estimated using the diffusion coefficient of acetone in water at infinite dilution and equation 1. The average diffusion coefficient is thus approximately 7×10^{-6} cm^2/s; again, heterogeneity and concentration dependence are neglected for the qualitative analysis used here.

Similarly to the volume change kinetics of HPC gel P46 observed in response to changes in solution ionic strength, the convection of solution through the pores enhances the stimulus change during swelling and retards the stimulus change during shrinking of gel P46 in response to change in the solvent composition. In fact, the magnitudes of the diffusion coefficients for swelling, shrinking, and solute are about the same as for the ammonium sulfate case. This consistency further strengthens the conclusions that the upper bound on the response rate obtainable for responsive gels is the rate of stimulus transfer and the lower bound observable is the rate of network diffusion.

Response to Changes in pH. A pH-sensitive gel swells upon ionization of ionic groups attached to the polymer chains and shrinks upon neutralization of ionic group. The rate of stimulus change in pH-sensitive gels will depend upon the diffusion of ions through the solution boundary film, the rate of diffusion of ions in the gel to the ionizable sites and the rate of ion exchange reaction at the ionic site. According to the literature, it is expected that the ion exchange reaction at the ionic site will be much faster than the rates of diffusion of ions through the boundary film and the gel *(14,15,28)*.

Since HPC gel is nonionic and thus doesn't change its volume in response to pH changes, it could not be used to determine the upper bound of volume change kinetics of pH-sensitive gels. However, the studies on the salt stimulus indicate that the rate of ion transfer would set the upper bound of the volume change kinetics of pH-sensitive gels as the rate of stimulus change is similar in both the cases. It is expected that the upper bound in the pH-sensitive gels will also depend upon the rate of transport of ions into or out of the gel and the solution boundary layer. Since the resistance offered by the solution boundary layer can be minimized by efficiently stirring the solution, the upper bound of the volume change kinetics should depend simply upon the transport rate of ions inside the gel. Therefore, just like ionic strength-sensitive gels, the transport rate of ions inside the pH-sensitive gel will be given by the diffusion rate of ions slightly influenced by the convection of the solution through the interconnected pores. Thus the upper bound of the volume change kinetics of pH-sensitive gels is expected to be governed by the rate of stimulus change similar to ionic strength-sensitive gels. It can be described by a diffusion coefficient of the order of 10^{-5} cm^2/s to 10^{-6} cm^2/s.

The lower bound of volume change kinetics for pH-sensitive gels observed in non-porous gels is generally set by the rate of network motion *(14)*. Therefore, the upper bound of volume change kinetics possible for microporous pH-sensitive gels with interconnected pores is expected to be 50 to 100 times greater than observed for non-porous gels. However, it has been found that in some cases (e.g. unstirred, dilute solutions), the transport rate of ions through the solution boundary layer and inside the gel can become slow enough to become the rate limiting step *(14,15,28)*.

Conclusions

Microporous responsive gels can be made that have much faster volume change kinetics than non-porous gels. The rate of network diffusion, with the sample dimension as the characteristic length, sets the lower bound of response that can be observed for responsive gels, and is characterized by a diffusion coefficient on the order of 10^{-7} cm^2/s. The fastest possible rates of volume change are seen in microporous, stimulus-sensitive gels. Microporous, temperature-sensitive gels have the fastest response, over 1000 times faster than chemically similar but non-porous gels. In these gels, the characteristic network dimension is that of the pore walls, and response time is estimated to be well under a second for structures of 1 μm in dimension. It was shown that microstructures can be made in which convection is also much faster than the rate of stimuli change. Thus the upper-bound on response rates observed in responsive gels is the rate of stimulus transfer. Since the rate of heat transfer is about 100 times faster than the rate of diffusion of ions or solvent inside the gel, the upper bound on the rate of volume change of temperature sensitive gel is about 100 times greater than that of ionic strength or solvent sensitive gels.

Acknowledgments

R. Spontak aided with the electron microscopy of the gels.

Literature Cited

1. Gehrke, S. H. In *Responsive Gels: Volume Transitions II*; Dusek, K., Ed., Springer-Verlag: Berlin, 1993, pp. 81-144.
2. Chiarelli, P.; Umezawa, K.; De Rossi, D. In *Polymer Gels: Fundamentals and Biomedical Applications.* De Rossi, D.; Kajiwara, K.; Osada, Y.; Yamauchi, A., Eds.; Plenum Press: New York, 1991.
3. Cussler, E. L.; Stokar, M. R.; Varberg, J. E. *AIChE J.* **1984**, *30*, pp. 578-582.
4. Gehrke S. H.; Andrews G. P.; Cussler E. L. *Chem. Eng. Sci.*, **1986**, *41*, pp. 2153-2160.
5. Freitas, R. F. S.; Cussler, E. L. *Chem. Eng. Sci.* **1987**, *42*, pp. 97-103.
6. Huang, X.; Unno, H.; Akehata, T.; Hirasa, O. *J. Chem. Eng. Jpn.*, **1987**, *20*, pp. 123-128.
7. Ross, P. E. *Forbes* , *2/14/94*, pp. 67-68.
8. Kabra, B.G. *Synthesis, Structure, and Swelling of Microporous Polymer Gels;* Ph.D. Thesis; University of Cincinnati: Cincinnati, OH, 1993.
9. Lyu, L. H. *Dewatering Fine Coal Slurries By Gel Extraction*; M. S. Thesis; University of Cincinnati: Cincinnati, OH, 1990.
10. Gehrke, S. H.; Lyu, L. H.; Yang, M. C. *Polym. Prepr.* **1989**, *30*, pp. 482-483.
11. de Gennes, P. G. *Scaling Concepts in Polymer Physics*; Cornell University Press: Ithaca, NY, 1979.
12. Tanaka, T.; Hocker, L. O.; Benedek, G. B. *J. Chem. Phys.* **1973**, *59*, pp. 5151-5159.
13. Cussler, E. L. *Diffusion: Mass Transfer in Fluid Systems*; Cambridge University Press: Cambridge, 1984.
14. Gehrke, S. H.; Cussler, E. L. *Chem. Eng. Sci.*, **1989**, *44*, pp. 559-566.
15. Gehrke, S. H.; Agrawal, G.; Yang, M. C. In *Polyelectrolyte Gels,*; Harland, R. S.; Prud'homme, R. K., Eds.; ACS Symposium Series 480; American Chemical Society: Washington, D.C. pp. 212-237.
16. Kabra, B. G.; Gehrke, S. H. *Polymer Commun.*, **1991**, *32*, pp. 322-323.
17. Kabra, B. G.; Gehrke, S. H.; Akhtar, M. K. *Polymer*, **1992**, *33,* pp. 990-995.
18. Kabra, B. G.; Gehrke, S. H. *U. S. Patent Application* , 1994.
19. Harsh, D. C.; Gehrke, S. H. *J. Controlled Release*, **1991**, *17*, pp. 175-185.
20. Tanaka, T.; Fillmore, D. F. *J. Chem. Phys.*, **1979**, *70*, pp. 1214-1218.
21. Crank, J. *The Mathematics of Diffusion, 2nd Ed.*; Oxford University Press: London, 1975.
22. Carslaw, A. S.; Jaeger, J. C. *The Conduction of Heat in Solids, 2nd Ed.*; Oxford Clarendon Press: Oxford, 1959.
23. Kabra, B. G.; Gehrke, S. H.; Hwang, S. T.; Ritschel, W. *J. Appl. Polym. Sci.*, **1991**, *42*, pp. 2409-2416.
24. Hirotsu, S.; Hirokawa, Y.; Tanaka, T. *J. Chem. Phys,* **1987**, *87*, pp. 1392-1395.
25. Palasis, M.; Gehrke, S. H. *J. Controlled Release*, **1992**, *18*, pp. 1-11.
26. Meares, P. In *Diffusion in Polymers,* Crank, J.; Park, G. S., Eds.; Academic Press: London, 1968, pp. 373-428.
27. Akhtar, M. K. *Volume Change Kinetics of Responsive Polymer Gels;* M. S. Thesis; University of Cincinnati: Cincinnati, OH, 1990.
28. Helfferich, F. *J. Phys. Chem.*, **1965**, *69*, pp. 1178-1187.

RECEIVED July 27, 1994

APPLICATIONS

Chapter 7

Trends in the Development of Superabsorbent Polymers for Diapers

Fusayoshi Masuda

Research Division, Sanyo Chemical Industries Limited, Kyoto 605, Japan

Superabsorbent polymers have become an important component of diapers during the last 10 years. World-wide demand for these polymers is currently about 320,000 metric tons annually. In this paper, we report the results of a study relating diaper performance to the properties of the superabsorbent polymer. We found that absorbency under load and gel stability are important polymer properties to relate to diaper leakage and surface dryness. However, these properties appear to approach an upper limit related to the current chemical structure of cross-linked polyacrylates. Our data suggests a limit for absorbency under load of 35-40 mL/g.

Commercial production of superabsorbent polymer began in Japan in 1978, for use in feminine napkins. Although the superabsorbents used today are based on poly(acrylic acid), this early superabsorbent was a crosslinked starch-g-polyacrylate (1), an improved version of the uncrosslinked starch graft polymer developed in the 1970's by M. O. Weaver and her associates at the Northern Regional Research Laboratory of the United States Department of Agriculture (2). After further development (3), superabsorbent polymer was used in baby diapers in Germany and France in 1980. These first diapers used only a small quantity of superabsorbent polymer, about 1-2 g per diaper, and the polymer was considered only a supplement to fluff pulp which furnished most of the absorbency. In 1983, a thinner diaper using 4-5 g polymer and less fluff was marketed in Japan. This was followed shortly by the introduction of thinner superabsorbent diapers in other Asian countries, the United States and Europe. More recently, diapers have become even thinner, relying on greater amounts of superabsorbent polymer. Specialty markets for superabsorbents have also developed in agriculture, sealants, air-fresheners and toys.

0097–6156/94/0573–0088$08.00/0
© 1994 American Chemical Society

Demand for Superabsorbent Polymers

Since the introduction of superabsorbent diapers in Japan in 1983, demand for superabsorbent polymer has increased rapidly, Figure 1. The demand for superabsorbent polymer in Japan in 1992 was about 32,000 metric tons. The polymer was used mostly in baby diapers, feminine napkins and adult diapers. Baby diapers account for the bulk of the demand. Demand for superabsorbents in feminine napkins is fairly stable at about 1,000 metric tons due to a mature market in Japan. The demand for superabsorbent in adult diapers has been gradually increasing in Japan. Currently, about one-third of the market for adult diapers uses superabsorbent polymer.

World-wide, the demand for superabsorbent polymer has increased dramatically since 1986, and currently amounts to about 280,000 metric tons per year (Figure 2). The largest consumption is in the the United States (130,000 metric tons per year), followed by Europe at 100,000 metric tons and Asia and Oceania at 20,000 metric tons. Based on the trend, we expect demand for superabsorbent polymers to reach 350-400,000 metric tons per year within a few years.

Application Methods in Baby Diapers

Superabsorbent polymer is added to baby diapers in basically two ways: layered or blended. The layered application is commonly adopted by Japanese diaper manufacturers. In this method, powdered superabsorbent polymer first is scattered onto a layer of fluff pulp. The fluff is then folded, so that the polymer is located in a centralized layer in the absorbent structure. This structure is covered with a non-woven fabric layer. In the blended application, the superabsorbent polymer first is mixed homogeneously with the fluff pulp. Then the mixture is laid down to give the absorbent structure, which is subsequently covered with a non-woven fabric. This method typically is adopted by American diaper makers.

In either method, containment of the powdered polymer within the loose, porous structure of the diaper is a concern. A recent development in Japan is the use of thermally bondable fibers within the absorbent structure to help fix the superabsorbent in place. In this method, some of the fluff pulp is replaced with thermally bondable fibers. The resulting absorbent core is heat-cured to give additional structural stability.

The use of superabsorbents has allowed thinner and lighter diapers to be made. The first superabsorbent diapers contained 55-60 g of fluff in addition to 1-2 g of polymer. Currently, diapers with less leakage use about 7 g of polymer per diaper and only 30-35 g of fluff.

Important Properties for Good Diapers

The first superabsorbent diapers were characterized by having improved surface dryness after absorbing urine. Reduced incidence of urine leaks has been added as an important performance parameter for current diapers.

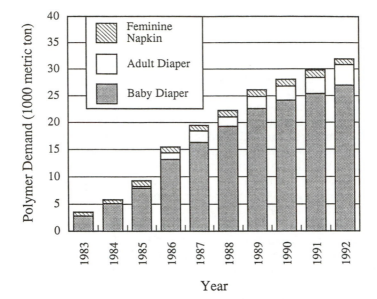

Figure 1. Demand for superabsorbent polymer in Japan, for feminine napkins, adult diapers and baby diapers.

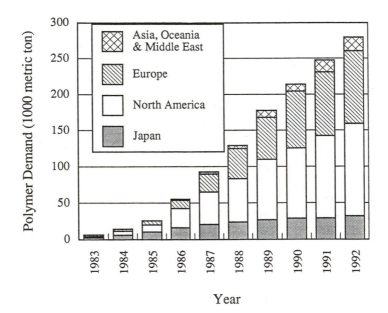

Figure 2. Demand for superabsorbent polymers worldwide, by geographical area.

Both are now considered as basic functions for modern diapers. A important problem in the design of new superabsorbent diapers is relating the physical properties of the superabsorbent polymer to the performance of the diapers containing the polymer. In order to clarify the relationship between diaper needs and polymer properties, a coefficient study was done (4).

First, we synthesized 30 samples of superabsorbent polymer having different combinations of properties. Then the polymers were incorporated into a standard diaper model, using the layered application method described earlier. Diaper dryness was determined by measuring the moisture content of the diaper surface. Leakage data was accumulated in actual diaper usage tests. The polymers alone were evaluated with 40 different test methods, including free-absorbency, absorbing rate, absorbency under load (AUL) and gel-strength. Figures 3-6 show the important results of this study of the relationship between diaper performance and polymer properties.

The free absorbency of the superabsorbent polymer was determined by measuring the increase in mass of one gram of polymer contained in a sealed tea-bag, after immersion in saline solution for 60 minutes. We found no relationship between diaper dryness and the free absorbency of the polymer. This is despite the widespread popularity of the free absorbency test and claims that high free absorbency yields a better diaper.

The absorbency under load (AUL) of the polymer was determined by measuring the volumetric take-up of saline into one gram of polymer that was spread on one square centimeter of filter-glass and compressed under a 20-gram weight. The saline solution was in contact with the polymer, through the filter-glass, for 60 minutes. Figure 3 shows the relationship of AUL to diaper dryness. We found a good relationship between higher AUL and improved diaper dryness.

Diaper dryness was also improved for polymers having higher re-absorbency of the sheared gel for saline solution. In this test, the polymer is first swollen with ten times its mass of saline solution and then mechanically sheared. The sheared gel is then put into a tea-bag and immersed in saline solution. The increase in mass is measured and the reabsorbency is calculated per gram of dry polymer. The relationship is shown in Figure 4. Another good relationship was found between diaper dryness and the elasticity modulus of the swollen gel. Elasticity modulus was measured on gel swollen with 40 times its mass of saline solution.

Diaper leakage was determined from actual use-testing of the model diapers. Of the 40 measures of performance of superabsorbent polymers that we performed, only the stability of the gel to shear could be related to diaper leakage. We found that polymers that were more stable to shear yielded diapers with less leakage. This correlation is shown in Figure 6. Gel-stability to shear was determined by measuring the elastic modulus of the polymer gel, swollen to 40 times its mass of saline solution, both

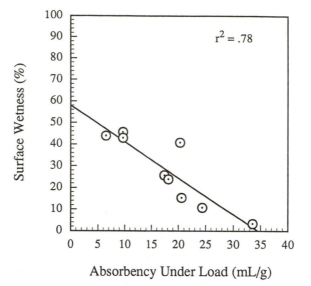

Figure 3. Correlation of surface wetness of diapers to the absorbency under load of the superabsorbent polymer used in the diaper.

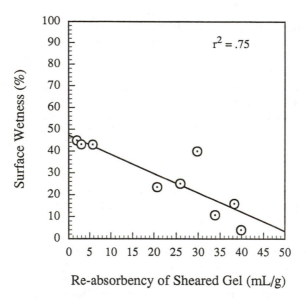

Figure 4. Correlation of surface wetness of diapers to the re-absorbency of liquid by the sheared gel.

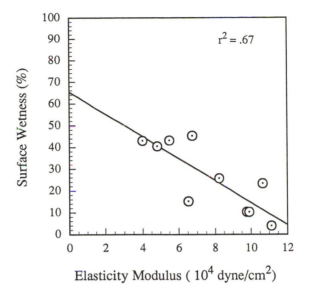

Figure 5. Correlation of surface wetness of diapers to the elasticity modulus of the swollen superabsorbent polymer.

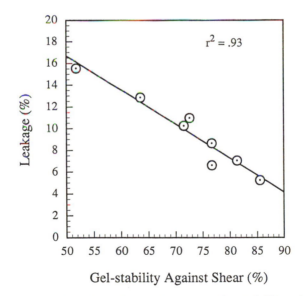

Figure 6. Correlation of leakage from diapers to the stability of the swollen gel to shear.

before and after mechanical shearing. The fraction of the original elastic modulus retained by the sheared gel is expressed as the percent stability.

The results of our study are summarized in Table I. Diaper dryness is closely correlated with three polymer tests: absorbency under load, reabsorbency of sheared gel and the elasticity modulus of the swollen gel. A balance of these three factors appears important for improved diaper dryness. Diaper leakage was closely correlated to the stability of gel to shearing. The scientific meaning of these four tests has not yet been fully investigated and more study is warranted. However, we think that superabsorbent polymer for use in baby diapers should have higher gel strength and an ability to maintain the absorption capability against shearing because the physical properties of the swollen gel are influenced by the shear that accompanies the movement of a baby while the diaper is in use.

Table I. Superabsorbent polymer functions related to the performance of diapers.

Diaper Needs	Superabsorbent Polymer Functions	Correlation Coefficient
Dryness	Absorbency Under Load	0.78
	Re-absorbency of Sheared Gel	0.75
	Elasticity Modulus	0.67
Less Leakage	Gel Stability against Shear	0.93

Trends in the Properties of Superabsorbent Polymers

With our new insight into four polymer properties that are important for dryness and less leakage, we wish to review the historical trends in these properties. We measured the free absorbency and the absorbency under load for a number of commercially available superabsorbent polymers. As shown in Figure 7, older samples had higher free absorbency and lower absorbency under load. More recently, the AUL has increased to about 30 mL/g while free absorbency has dropped to about 50 mL/g. Based on extrapolation of Figure 7, we think that AUL may be increased to 35-40 mL/g using conventional, or existing, technology. Because of its demonstrated importance to diaper performance, it may be desirable to have AUL as high as 50 mL/g. We think a technical breakthrough is necessary to achieve this value, however. The historical trend of gel stability to shearing is shown in Figure 8. Older samples had poor stability to shear but this property has been gradually improved over time, with recent samples having 70-90 % retention of modulus after shearing.

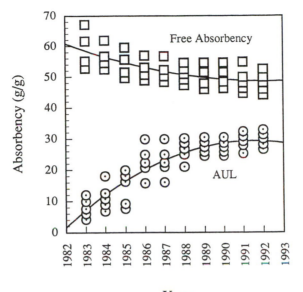

Figure 7. Historical trends of the absorbent properties of superabsorbent polymers.

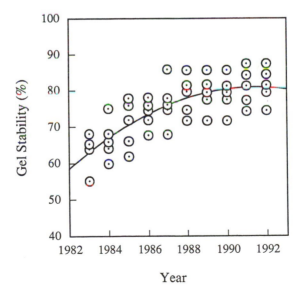

Figure 8. Historical trend of the stability of swollen gel to shear.

Trends in Process Control of Superabsorbent Polymers

Polyacrylate superabsorbent polymer has been in commercial production for ten years. The absorbency parameters are not the only features that have changed over this time. As shown in Figure 9, the level of residual acrylic acid has dropped from over 1000 ppm in 1983 to around 100 ppm in 1992. Figure 10 shows that the extractable polymer fraction of superabsorbent, comprising low molecular weight and slightly crosslinked polyacrylate, has also decreased over time. More stringent raw material specifications by the diaper producers contributed to the current superabsorbents having reduced amounts of fines (particles less than 50 microns) and a narrower particle size distribution (Figure 11).

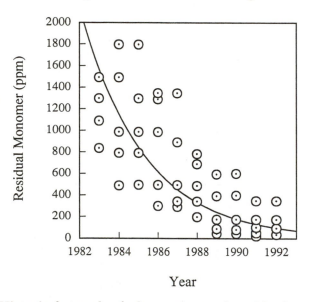

Figure 9. Historical trend of the content of residual monomer in superabsorbent polymers.

Conclusion

We have found that absorbency under load and stability of the gel against shear are important properties of superabsorbent polymers and relate strongly to diaper performance. However, these properties appear to be approaching a limit related to the current chemical structure of cross-linked polyacrylates. Because of the market requests for a thinner diaper, more superabsorbent polymer and less fluff is being incorporated into diapers. This approach limits the maximum amount of superabsorbent polymer in a diaper to about 10 g/piece, and this requires that the absorbency under load be increased. A target for AUL of 35-40 mL/g is probably achievable using current technology, but an AUL of about 50 mL/g is probably necessary to get a much thinner diaper. The future efforts

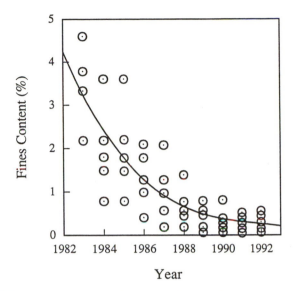

Figure 10. Historical trend of the content of extractable polymer in superabsorbent polymers.

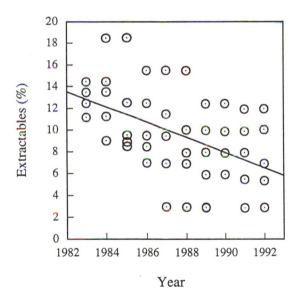

Figure 11. Historical trend of the content of fine particles (less than 50 μm) in granular superabsorbent polymers.

of manufacturers to improve the production and engineering of superabsorbents should make possible higher performance and even lower levels of residual monomer, extractables and fines in the product.

Literature Cited

1. Masuda, F.; Nakamura, A.; Nishida, K. US 4,076,663 (1978).
2. Weaver, M. O.; Bagley, E. B.; Fanta, G. F.; Doane, W. M. US 3,981,100 (1976).
3. Masuda, F. *Superabsorbent Polymers*, Ed. Japan Polymer Society, Kyoritsu Shuppann, 1987, 24-50.
4. Chambers, D. R.; Fowler, H. H.; Fujiura, Y.; Masuda, F. US 5,145,906 (1992).

RECEIVED July 27, 1994

Chapter 8

Characterization of a New Superabsorbent Polymer Generation

Herbert Nagorski

Stockhausen GmbH, D–47705 Krefeld, Germany

Choosing a superabsorbent polymer that will perform optimally in the absorbent core of a diaper or other personal care product is a difficult design problem for diaper manufacturers. Equally challenging is the understanding of which property of the superabsorbent polymer must be improved to achieve the optimal performance. Laboratory evaluations of absorbent pads containing new superabsorbents show that superabsorbents with increased absorbency under load, increased swelling pressure and increased suction power are beneficial in thinner pads having less cellulose fluff and more superabsorbent polymer.

When superabsorbent polymers (SAP) were first introduced into the absorbent cores of personal care articles, only two properties were considered to be of significance: (1) total absorption and (2) retention capacity. The concentration of superabsorbent polymer in the absorbent core of these early hygiene articles was fairly low (Fluff/SAP ratio = 10). The only task of the SAP was to absorb and retain the liquid absorbed by the fluff-fiber matrix. Thus, a superabsorbent polymer with high retention capacity was generally obtained by manufacture of a lightly crosslinked, partially neutralized poly(acrylic acid). High gel stability, needed to prevent gel blocking, was not a design parameter for superabsorbent polymers at that time. These materials may be called first generation superabsorbent polymers.

A few years later, the construction techniques for absorbent cores changed to use a targeted placement of superabsorbent polymer. In pads with local areas of high concentration of superabsorbent polymer, the lightly crosslinked superabsorbents failed due to the gel blocking effect. A new superabsorbent polymer property, absorption under load (AUL), was identified to be important for the performance of absorbent pads (1).

0097–6156/94/0573–0099$08.00/0
© 1994 American Chemical Society

Absorbency Under Load

Absorbency under load (AUL) measures the ability of a polymer to absorb fluid while under a static load and can be considered as a measurement of gel stability or gel strength. In early work, a comparison of AUL values with more scientific measurements of gel strength had shown a good correlation. A high value of AUL correlates to a high gel strength, and a high gel strength helps maintain a more porous gel mass during hydration, even at high concentrations of superabsorbent polymer. This improves fluid acquisition and liquid distribution in absorbent cores.

An improved polymer gel strength typically has been achieved by increasing the degree of crosslinking. When the AUL value (which is a function of the degree of crosslinking) is increased in this way, the retention capacity (which is a reciprocal function of the degree of crosslinking) is reduced at the same time, Figure 1.

A second generation superabsorbent polymer with an increased gel strength typically is characterized by a retention capacity of about 28-30 g/g and AUL (0.3 psi) values of 20 g/g or greater compared to a retention capacity and AUL of 37 g/g and 7 g/g for a first generation polymer. When the SAP is placed into special target zones of the diaper, this reduction of the retention capacity is still acceptable.

Most recently, the developments in diaper constructions clearly showed a need for superabsorbent polymers with even higher gel strength. A further improvement of the AUL by increasing the degree of crosslinking (see the dotted line in Figure 1) is undesirable, because this decreases the retention capacity too much.

Fluff Reduction and the Need for Higher AUL

The need for superabsorbent polymers with very high gel strength is necessitated by the reduction of fluff pulp in modern absorbent core constructions. The reasons for this fluff reduction are based on environmental and economic aspects (2,3). Fluff reduction offers thinner articles with better fit and more comfort. In addition, these thinner diapers use a lower volume of raw materials, which means less solid waste and lower packaging and shipping costs. But the reduction of fluff can lead to a reduced absorption and retention capacity of the pad. Therefore, the diaper producer may need to add additional quantities of superabsorbent polymer to maintain the absorption performance. Superabsorbent polymer concentrations of up to 50 % and even more have already been seen in the marketplace. For absorbent pads with such a high concentration of superabsorbent polymer , a third generation of superabsorbent polymer has been developed. This product is characterized as a surface cross-linked particle, Figure 2.

The basis of this superabsorbent polymer is a lightly crosslinked, superabsorbent polymer such as those of the first generation of superabsorbent polymers. By application and reaction of a second crosslinker on the surface of these particles, the particle becomes more highly crosslinked on the surface. In other words, the surface crosslinked superabsorbent particle contains a discontinuous network. This reaction type offers new and more numerous product variations because of the possibility of modifying the two different, crosslinked zones independently (4). The resulting advantage in properties is shown in Figure 3.

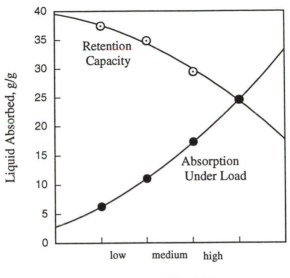

Figure 1. Retention capacity and absorption under load as a function of crosslink density of the superabsorbent polymer.

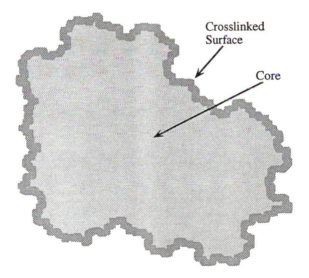

Figure 2. Third generation superabsorbent polymer particle with a crosslinked surface.

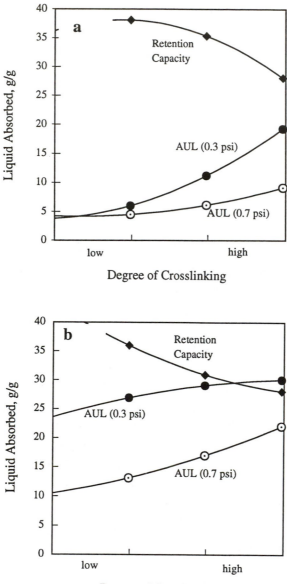

Figure 3. Retention capacity and absorption under load at two different pressures for (a) first and second generation and (b) third generation superabsorbents, as a function of degree of crosslinking.

As previously shown, the retention capacity decreased from 38 to 28 g/g during the transition from first to second generation superabsorbent polymers as the AUL(0.3 psi) increased. The AUL value determined under an increased static pressure of 0.7 psi was very low for both of these superabsorbent polymers. It is about 5 g/g for a first generation superabsorbent polymer and still less than 10 g/g for a second generation superabsorbent polymer.

Within the group of the third generation superabsorbent polymers, products with retention capacities of 28-37 g/g have been synthesized by increasing the degree of crosslinking of both crosslinked zones. The AUL values at the lowest pressure (0.3 psi) have been found to be 26 g/g; this is larger than the AUL (0.3 psi) value of second generation superabsorbent polymer. For the third generation superabsorbent polymers, an increase in the degree of crosslinking has no measurable effect on the AUL (0.3 psi) value. At the same time, the AUL measured under the static load of 0.7 psi shows a clear dependence on the degree of crosslinking. The crosslinking of the surface increased the AUL (0.7 psi) from about 12 to 20-22 g/g. A comparison shows that the AUL (0.7 psi) of a third generation superabsorbent polymer can be as high as the AUL (0.3 psi) of a second generation superabsorbent polymer. Thus the new generation of superabsorbent polymers show a much increased gel strength and may be able to work at very high concentrations in absorbent core constructions.

To test this assumption, a series of measurements has been performed in which the basis weight (areal mass distribution of the polymer layer) in the AUL test device was varied between 0.02 and 0.04 g/cm^2. (This range is typically achieved in modern diapers with thin absorbent cores.) Figure 4 shows that the AUL (0.3 psi) of a second generation superabsorbent polymer drops from about 19 to 11 g/g when the basis weight is increased in the above mentioned range. In comparison, a third generation superabsorbent polymer shows almost an unchanged AUL (0.3 psi) within this concentration range: the AUL value drops only from 32 to 30 g/g. This shows that the third generation superabsorbent polymers are able to work in very high concentrations.

Influence of the Absorption Speed of Superabsorbents on Pad Performance

It is important to know whether total absorption capacity, retention capacity, and absorption under load are the only properties of SAP that influence pad performance. We must keep in mind that by the reduction of fluff in modern absorbent core constructions, the superabsorbent polymer has to take over more of the functions of the fluff pulp. The main functions that must be considered for an absorbent core are: (1) absorption, (2) retention, (3) acquisition, (4) rewet behavior, (5) wicking and liquid distribution and (6) leakage behavior. In thinner diapers, the fluff contributes little to absorption and retention; these can be ideally achieved by the SAP. Due to its fiber structure, fluff pulp generally offers a fast acquisition of fluids as well as excellent wicking and liquid distribution. In these aspects, the superabsorbent polymer often fails.

Additional polymer properties that have been identified to have a significant influence on pad performance are speed of absorption, liquid distribution and suction power. While the speed of absorption may be measured in a number of ways, a simple

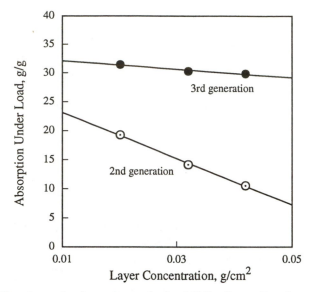

Figure 4. The change in absorption under load (0.3 psi) as a function of the areal concentration of superabsorbent polymer in the test device for different superabsorbent polymer generations.

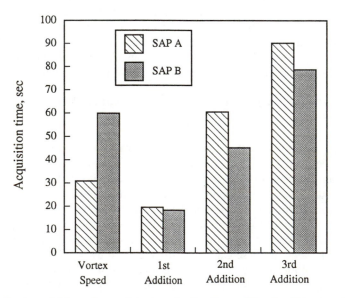

Figure 5. Acquisition time of test pads for three 60-mL aliquots of fluid, compared to the vortex speed of the polymers alone; (A) vortex time = 24 sec, (B) vortex time = 60 sec. Both samples have a retention capacity of 28 g/g.

measurement technique is called the vortex test, which yields the time to absorb a given volume of liquid. In test pads made in the Stockhausen laboratory, we have measured the influence of vortex speed on the acquisition time of several liquid additions to these pads. For this test, a second generation superabsorbent polymer A and a third generation superabsorbent polymer B were chosen that showed differences in the vortex speed (superabsorbent polymer A, t = 24 sec; superabsorbent polymer B, t = 60 sec). As Figure 5 clearly indicates, the superabsorbent polymer B with the *slower* vortex speed offers a *faster* uptake of the liquid into the pad.

This is a result of the change in the structure of the pad matrix containing the faster absorbing polymer A. As a result of the change in the pad matrix, capillarity is changed so that further liquid additions are not wicked away from the loading point very quickly. The superabsorbent polymer B does not absorb the liquid as fast, so that the pad structure is not significantly changed. Then, a further liquid addition wicks away much faster because capillarity has been maintained. The absorption speed, determined as absorption under load as a function of time, shows a much better correlation to the acquisition time of absorbent core constructions, Figure 6.

Whereas the second generation superabsorbent polymer A shows an AUL uptake of about 5 g/g after 1 min, the AUL value of a third generation superabsorbent polymer B is already 19 g/g within the same time interval. Furthermore, the absorption of the third generation superabsorbent polymer is complete after 5-6 min. The AUL of the second generation superabsorbent polymer A, at ten minutes, is well below its final value and continues to increase at a constant rate. Other superabsorbent polymer properties influencing wicking and liquid distribution are suction power and swelling pressure.

These properties both increase in the third generation superabsorbent polymers compared to first and second generation superabsorbent polymers, Figure 7. For a third generation superabsorbent polymer the suction power is twice as high as for a second generation superabsorbent polymer. The swelling pressure is even tripled.

Influence of Suction Power and Swelling Pressure on Pad Performance

To show the effect of suction power, we prepared simulated test pads. A fluff layer on top was designed to act as an acquisition layer. A high concentration layer of superabsorbent polymer was placed near the bottom to serve as the storage area of this core construction. Because such a pad construction usually shows a good acquisition rate but poor rewet behavior, we also placed a layer with a low concentration of superabsorbent polymer near the top.

Figure 8 shows a comparison of the performance of different polymers in this core design. We observed the shortest acquisition times and the lowest rewet values in the pads with the third generation superabsorbent polymer, which has the highest suction power. We believe that the high suction power of the third generation superabsorbent polymer attracts the interstitial water that was initially absorbed by the fluff fiber matrix. This dries the fiber matrix and maintains its capillarity. Thus, the next liquid addition can be absorbed much faster, and the rewet behavior is improved.

Most recent diaper designs have extremely thin absorbent cores. Under normal conditions, in these articles, the absorbent core would increase in thickness during the

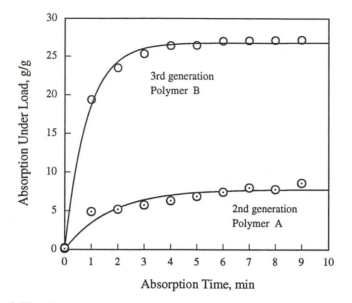

Figure 6. The absorption speeds of SAP A and SAP B while under a load of 0.3 psi. The samples are the same as in Figure 5.

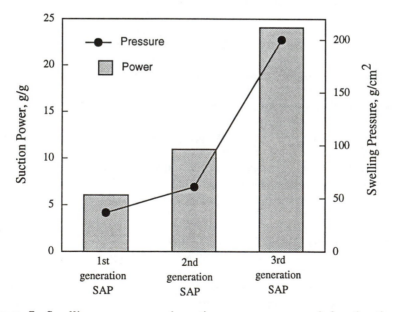

Figure 7. Swelling pressure and suction power compared for the three generations of superabsorbent polymers.

absorption process, because the pad would otherwise not be able to absorb all the discharged liquid. Usually, the thin pads are compressed to give a high density. This destroys the resiliency of the fluff fiber matrix, reduces capillarity and therefore, decreases the ability of the fluff matrix to move and store liquid. Therefore, the superabsorbent polymer must increase the pad thickness as it swells. To understand these effects, swelling pressure and swelling height are measured to estimate the ability of the polymer to increase the thickness of the pad as the polymer swells. The swelling height of three superabsorbent polymers having different swelling pressures was measured as function of time, Figure 9.

A second generation superabsorbent polymer with a typical swelling pressure of about 50 g/cm^2 yields a swelling height of about 3 mm after 20 minutes. For different third generation superabsorbent polymers, the swelling pressure was adjusted by surface crosslinking. Swelling heights of 9 mm and 10 mm were measured after 20 minutes of swelling for polymers with swelling pressures of 90 g/cm^2 and 135 g/cm^2, respectively. Thus the swelling height is clearly a function of the swelling pressure.

Up to this point, we have shown that properties other than absorption, retention and AUL are important to pad performance. To get the best picture of how superabsorbent polymers work in an absorbent core, we established a test system that includes the determination of properties of the superabsorbent polymer itself, of simulated test pads and of the personal care articles made by hygiene article producers, because it is important that laboratory test results correlate to the results during use.

To choose or develop the best performing polymer for a particular application, there is a need for a clear definition on the requirements. In order to accomplish this, a close relationship between the polymer producer and the producer of the personal articles must be established. The selection and development of superabsorbent polymers can not be accomplished by looking exclusively to superabsorbent polymer properties.

Experimental Section

Retention Capacity. The retention capacity is determined by weighing 0.2 g of the superabsorbent polymer in a teabag (6 x 6 cm). The sealed teabag is placed in 0.9% sodium chloride aqueous solution for 30 minutes and allowed to swell. The soaked teabag is removed from the saline solution and allowed to drip dry for 5 minutes before centrifugation at a force of 250 G for 5 minutes. The retention capacity of the superabsorbent polymer is the weight of liquid in the teabag remaining after centrifugation divided by the initial weight of the superabsorbent polymer.

Absorbency Under Load (AUL). A porous filter plate is placed in a petri dish and 0.9 % sodium chloride solution is added so that the liquid level is equal to the top of the filter plate. A filter paper is placed on the filter plate and allowed to thoroughly wet with the saline solution. Superabsorbent polymer (0.9 grams) is carefully scattered onto the filter screen of the test device (a Plexiglas cylinder with 400 mesh stainless steel cloth in the bottom: cylinder diameter = 60 mm, height = 50 mm). A piston assembly, including additional weight to achieve a load of 0.3 psi (or 0.7 psi), is placed on top of the superabsorbent polymer. After weighing the assembled device, it is placed on the

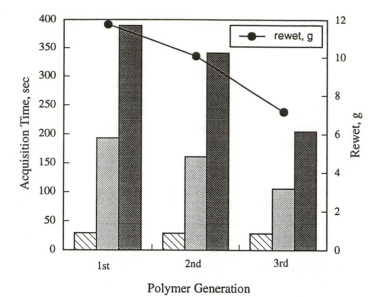

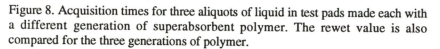

Figure 8. Acquisition times for three aliquots of liquid in test pads made each with a different generation of superabsorbent polymer. The rewet value is also compared for the three generations of polymer.

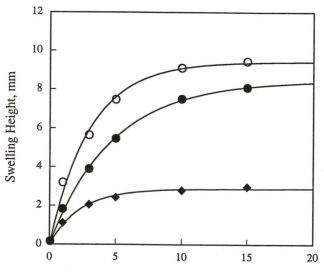

Figure 9. Swelling height of superabsorbent polymers as a function of time for samples differing in their swelling pressure; O p=135 g/cm^2, ● p=90 g/cm^2, ◆ p=50 g/cm^2.

filter plate, and absorption is allowed for 1 hour. After 1 hr, the entire device is re-weighed and the absorbency under load is calculated by the following formula:

$$AUL = \frac{\text{(Mass of cylinder group after suction - mass of cylinder group dry)}}{\text{(initial sample mass of the superabsorbent)}}$$

Suction Power. The suction power of a superabsorbent polymer is determined in a manner similar to the AUL measurement. The difference in this analysis technique is that liquid is not freely supplied as in the AUL test, but is supplied instead by a fluff pad that is 50% saturated with 0.9% saline solution, Figure 10. In this case, the superabsorbent polymer must absorb the liquid against the capillary tension of the partially saturated fluff pulp matrix.

Swelling Pressure. The superabsorbent polymer (0.5 g) is placed in a dry plastic cylinder. Saline solution (30 mL) is added to the cylinder, then the measuring device used to determine the swelling force is placed in the cylinder so that a fixed open volume of 14 cm^3 is achieved. As the polymer swells against the piston, the force generated is determined by the measuring device, Figure 11.

Vortex Test. Fifty mL of a 0.9% saline solution is added to a glass beaker (100 mL) and stirred with a stirring rod at 600 rpm. Exactly 2.00 g of the superabsorbent polymer is carefully introduced into the vortex. At the same time a stop watch is started. When the the vortex disappears, the watch is stopped and the time is recorded as the vortex speed.

Test Pad Preparation. Fluff pulp to produce simulated test pads is defibrated in Kamas H 01 laboratory hammer mill. Pads are formed with 30 g of defibrated fluff that is fed into a stationary pad lay-down system (30 x 30 cm). The superabsorbent polymer is added simultaneously for the preparation of homogenous blends or is added at a later time to create layer constructions. A pneumatic press is used to compress the pads to the desired density.

Pad-Rewet. To determine the pad-rewet, a 12 x 30 cm pad is prepared according to the description above. The pad is placed into a shaped test device as shown in Figure 12. The pad is placed between the inner plastic body and the PE-foil and a mass of 9 kg is added on top of the plastic body. After adding three 80-mL portions of 0.9 % saline solution at 20 min intervals thru the addition column, the pad is removed from the device and placed on a table. Three stacks of filter paper are placed on the pad surface, under a load of 20 g/cm^2. After 10 min, the filter paper is reweighed and the uptake of liquid is recorded as rewet. With each portion of saline solution, the time to absorb the liquid by the pad matrix is recorded.

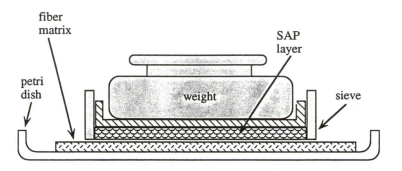

Figure 10. Schematic of the device used for measuring the suction power of a superabsorbent polymer against the capillary pressure of a fiber matrix.

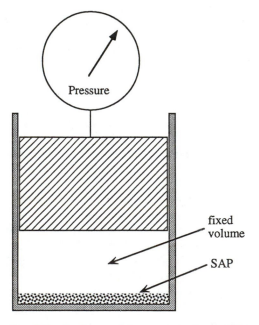

Figure 11. Schematic of the device used to measure the swelling pressure of a superabsorbent polymer that is confined to a fixed volume as it swells.

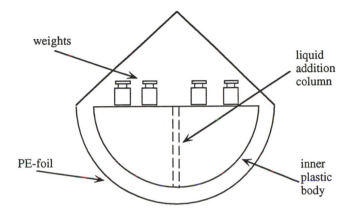

Figure 12. Schematic drawing of the device used to measure the rewet performance of pads containing superabsorbent polymer.

References

1. F. Masuda, "The Concept of Superabsorbent Polymer", Paper No. 13 , Pira Conference, Pira, Randalls Road, Leatherhead, Surrey KT22 7RU, England, December 1987
2. Ian Cheyne, "Ultrathin Absorbent Composites using Superabsorbent Fibers", Insight-Conference 1993, Toronto,Canada, October 1993
3. Andrew Urban, "Diaper Trends - Historical and Future Perspectives", IDEA 1992 Book of Papers, INDA Association of the Nonwoven Fabrics Industry, 1001 Winstead Drive, Suite 460, Cary, North Carolina 27513
4. DE 40 207 803, K. Dahmen, R. Mertens, Chemische Fabrik Stockhausen, 1990

RECEIVED July 27, 1994

Chapter 9

Preparation and Application of High-Performance Superabsorbent Polymers

Tadao Shimomura[1] and Takashi Namba[2]

[1]Himeji Research Laboratory, Nippon Shokubai Company Limited, 992−1 Aboshi-ku, Himeji, Hyogo 671−12, Japan
[2]Central Research Laboratory, Nippon Shokubai Company Limited, 5−8 Nishi Otabi-cho, Suita, Osaka 564, Japan

Since the first superabsorbent polymer (SAP) was reported by the U. S. Department of Agriculture (*1*), SAP research and development has focused on its extraordinary high water absorbency and its applications. The most remarkable success has been in hygiene applications such as disposable diapers and sanitary napkins; over 80% of SAP is currently consumed in these applications. Because the SAP market in hygiene applications has been extremely competitive, SAP research and development have been greatly accelerated and have unveiled many unique properties of SAP other than high water absorbency. Many new applications have been investigated through such active research and development, and SAPs have already been used effectively in some new applications. As a result, their consumption has steadily increased year by year. Furthermore, there are still many potential applications under development. However, at the same time, the research and development showed that the most common SAPs, those of the poly(acrylic acid)-type, are not always suitable for new applications since they are especially designed for hygiene applications and are supplied in powder form. In order to fit the properties of SAP into the new applications, physical and chemical modifications are sometimes required. In this chapter, some new high performance superabsorbent polymers and their applications will be described.

Important Basic Properties of Superabsorbent Polymers

Absorption of Aqueous Solution. The absorbency of a SAP must be measured under various conditions because it can limit the application areas of the material. For example, poly(acrylic acid)-type SAP cannot be used for applications where the SAP will be exposed to sea-water because the absorbency of poly(acrylic acid)-type SAP dramatically decreases due to its poor absorbency stability against polyvalent cations.

0097−6156/94/0573−0112$08.00/0
© 1994 American Chemical Society

The absorbency of a SAP has been defined by Flory as the swelling ratio, q_m, of the crosslinked polyelectrolyte network system, equation 1 (2).

$$q_m^{5/3} = \frac{(i/2v_u S^{*1/2})^2 + (1/2 - \chi_1)/v_1}{V_0/v_e} \tag{1}$$

where q_m is the swelling ratio of the network at equilibrium, i is the degree of ionization multiplied by the valency of the fixed charge in the network, v_u is the molar volume of a structure unit, S^* is the ionic strength of external solution, χ_1 is the interaction parameter between the network and the solvent, v_1 is the molar volume of the solvent, v_e is the effective number of chains in the network, and V_0 is the volume of the unswollen network.

In equation 1, the first and second terms in the right side represent the contribution of ionic charges in the network and in the solution, and the contribution of the affinity between the polymer network and the solvent, respectively. The term on the bottom of the right side represents the crosslink density of the network. Equation 1 explains the absorbing behavior of a SAP very well and can conveniently be used for designing a SAP since it gives basic ideas about what the main chain structure and the crosslink density of SAP should be for a particular application.

Figure 1 shows the absorbency dependence of a poly(acrylic acid)-type SAP on the concentration of NaCl and $CaCl_2$ aqueous solutions. The absorbency is measured by the tea-bag method after 30 min soaking (4). The absorbency decreases as the concentration of the salts increases; this is explained by the increase of $S^{*1/2}$ in equation 1.

In the case of $CaCl_2$ solution, the absorbency of poly(acrylic acid)-type SAP is also dependent on the soaking time. Figure 2 shows the time dependence of the absorbency of various crosslinked hydrophilic polymers in $CaCl_2$ solution measured by the tea-bag method. The initial absorbency and that after 1000 minutes of soaking are shown in Table I. The absorbency of the poly(acrylic acid)-type SAP dramatically decreases and finally, it almost completely loses its absorbency. This is because of the formation of additional crosslinks by the irreversible ion exchange between calcium ion and sodium ion on carboxyl groups, in other words because of the increase of V_0/v_e in equation 1.

Figure 2 also clearly shows how functional groups affect the absorbency stability of SAP against polyvalent cations. Crosslinked PAMPS showed absorbency decrease similar to that of the poly(acrylic acid)-type, but it maintained 12% of its initial absorbency after 1000 minutes of soaking. This is because of the reversible ion exchange on the stronger acid group than carboxylic acid; thus additional crosslinks are not permanent and the absorbency reaches a certain equilibrium. Crosslinked PTMAEA-Cl showed better absorbency stability than the poly(acrylic acid)-type and maintained 40% absorbency since the cationic group would not exchange any cations.

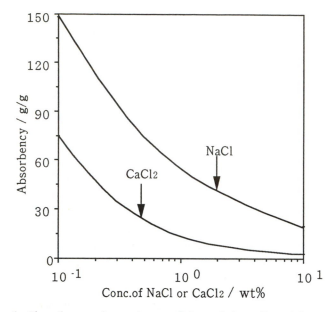

Fig. 1 Absorbency dependence of the poly(acrylic acid)-type SAP on the conc. of NaCl and CaCl₂ solution. (Adapted from ref. 3)

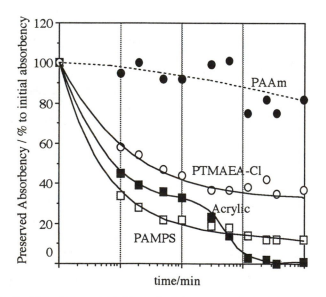

Fig. 2 Absorbency stability of various crosslinked hydrophilic polymers in 0.05 wt% CaCl₂ solution.

Table I. Absorbency of various crosslinked hydrophilic polymers in CaCl$_2$ solution

Type of Polymer	Initial Absorbency g/g	Absorbency at 1000 min g/g
Poly(acrylic acid)-type [a]	275	2
PAMPS [b]	394	46
PTMAEA-Cl [c]	329	122
PAAm [d]	11	9

(a) Partially neutralized crosslinked poly(acrylic acid)
(b) Crosslinked poly(sodium acrylamide-2-methyl-propane sulfonate)
(c) Crosslinked poly(chloro-trimethylaminoethyl-acrylate)
(d) Crosslinked poly(acrylamide)

Crosslinked PAAm had the best absorbency stability and maintained over 80% of its initial absorbency because the nonionic group is far less sensitive to ions than the ionic functional groups. (Since the absorbency of PAAm was so low, the absorbency change less than 20 % might be the experimental error.) However, it should be pointed out that PAAm had the least initial absorbency.

If the absorbency stability is the only consideration in high performance SAP development, PAAm or the nonionic SAP is essentially the best. However, in real SAP applications, many other factors should be considered, and this will be discussed in the following section.

Moisture Absorption. Figure 3 shows the moisture absorption of a poly(acrylic acid)-type SAP and silica gel. In this experiment, the weight increase of 1 g of the poly(acrylic acid)-type SAP powder or silica gel under various relative humidities was measured. The moisture absorption of SAP was much higher than that of silica gel at high relative humidity. The steeper slope at high relative humidity shows that the SAP is more sensitive to the humidity change, and more effectively absorbs and releases moisture than silica gel. Thus, the SAP could be a better moisture controlling material than silica gel.

Ammonia Absorption. Figure 4 shows the ammonia absorption of poly(acrylic acid)-type SAP that is generally used for hygiene applications. In this experiment, 1 g of the SAP was put into a tea-bag and hung in a 1-L closed plastic bottle where 2 mL of 0.1% ammonia solution was contained, and ammonia vapor concentration in the bottle was measured as a function of time. The SAP absorbs ammonia and the concentration of ammonia is dramatically decreased in a very short time since carboxylic acid groups in the SAP are partially neutralized and the residual acid groups can react with alkaline substances such as ammonia very effectively. This property is very beneficial in diapers because it prevents diaper rash by neutralizing ammonia in urine.

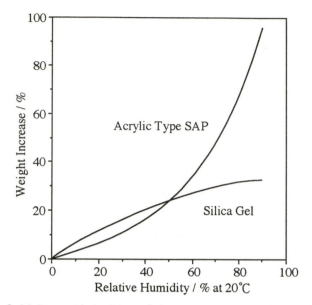

Fig. 3 Moisture absorption of the poly(acrylic acid)-type SAP and silica gel. (Adapted from ref. 3)

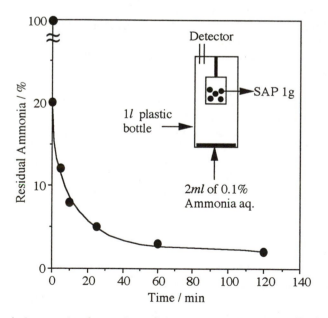

Fig. 4 Ammonia absorption of the poly(acrylic acid)-type SAP. (Adapted from ref. 3)

Water/Organic Solvent Mixture Absorption. Figure 5 shows the absorption of mixtures of water and organic solvent by a poly(acrylic acid) type-SAP, as measured by the tea-bag method (*4*) with 1 hr soaking. The absorbency is altered by the combination of water and organic solvents, and is also dependent on the ratio of water to organic solvent in the mixture. Note that the absorbency decreases sharply at a certain ratio. This is the phase transition of a gel, and the gel dramatically shrinks at this point. The phase transition can also be caused by changing other conditions such as temperature, pH, light, and electric field (*5*).

High Performance Superabsorbent Polymers

Absorbency Stable SAP. As shown in Figure 2, the absorbency stability of SAP depends on the type of functional groups present. If the absorbency stability is the only consideration in high performance SAP development, PAAm (the nonionic SAP) is essentially the best. However, in real SAP applications, many other factors should be considered such as absolute value of absorbency, safety, cost and productivity. The poor absorbency stability may even be considered as an advantage for some applications. For example, the volume of SAP gel waste will be dramatically decreased by contact with a liquid containing polyvalent cations, when the SAP has poor absorbency stability. Table II shows some important features of various SAPs that should be considered in order to develop a SAP for each specific application. The SAP whose performance and cost are the best-balanced will be used for the application.

Table II. Features of Various Superabsorbent Polymers

	Absorbency	Absorbency Stability	Monomer Cost	Others
Poly(acrylic acid) type (carboxylate)	High	Low	Low	High Safety High gel strength
PAMPS (sulfonate)	High	High	High	
PAAm (nonionic)	Low	High	Low	High Toxicity
PTMAEA-Cl (cationic)	High	High	High	Low gel strength

Superabsorbent Polymer Composites.

Water-Swelling Rubber. A water-swelling rubber is one of the most popular SAP composites and is used as a sealing material in construction applications. The SAP used for the water-swelling rubber is usually the absorbency-stable type because the rubber will be exposed to seawater, underground water, or cement bleeding water.

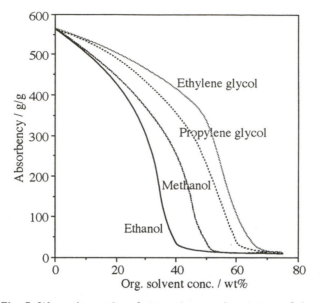

Fig. 5 Water/organic solvent mixture absorption of the poly(acrylic acid)-type SAP. (Adapted from ref. 3)

Fig. 6 A tunnel constructed with segments. (Reproduced with permission from ref. 3. Copyright 1991 T. Konno)

Basically, a water-swelling rubber can be prepared by blending SAP powder into rubber. However, this is not easily accomplished because of low interaction between the hydrophilic SAP powder and the hydrophobic rubber. Modification of the interface is usually required (*6-7*). Insufficient adhesion at the interface may cause the SAP powder to separate from the rubber and thereby hurt properties of the composite. In addition, the particle size of the SAP is important because the surface of water-swelling rubber will be rough if it is too large.

Figure 6 is a picture of a tunnel whose wall is made from concrete blocks. The water-swelling rubber is placed on all sides of each block, and fits together with the rubber on neighboring blocks. Once the rubber makes contact with water, it swells and fills the spaces between each block tightly, and prevents water from leaking into the tunnel. There are several examples of the segment tunnel construction for railways, subways, and highways in Japan, and it was also used in the Euro Tunnel under the Straits of Dover.

Water-Blocking Tape. Figure 7 shows the schematic cross sectional image of an optical communication cable containing water-blocking tapes. Water-blocking tapes are widely used in communication cables and power cables to prevent the cable system from being damaged by water intrusion when the outer cover of the cable is damaged.

Water-blocking tape is prepared by applying a dispersion of SAP onto a substrate. The dispersion consists of SAP powder and a binder such as styrene-butadiene-styrene (SBS) resin, which gives flexible binding between the SAP powder and the substrate (*8-9*). The substrate is usually a non-woven fabric, to give a flexible tape and keep a path of water to the SAP powder. Since the cable system will be exposed to seawater or underground water, the absorbency-stable type SAP is preferred for this application (*10*).

Figure 8 shows the length of intrusion of sea water into an optical communication cable containing water-blocking tape. The absorbency stable SAP completely stopped the intruding seawater in a shorter length than did the conventional SAP.

Web-type SAP. In instances when SAPs are used for hygiene applications, they are usually utilized in powder form. However, the powder form SAP has two major limitations: (1) inconvenience during handling and (2) instability in or on a substrate. For example, in dew prevention, SAP should be immobilized on walls and ceilings, but it is extremely hard to apply SAP powders on these surfaces directly. In order to overcome these two drawbacks, superabsorbent sheets or superabsorbent fibers have been developed.

Classical superabsorbent sheets are prepared by laminating the SAP powders between two pieces of paper (SAP-laminate sheet). SAP-laminate sheets dramatically improv SAP handling and are widely used as absorbent pads for meat and poultry to absorb juice or liquid exuded from them. However, they are not good enough to apply on vertical surfaces or ceilings because SAP powders in the sheet are placed so loosely that the powders or swollen gels can easily move. The loose binding also creates another problem in the performance of the sheet as a moisture-controlling device because the powders tend to agglomerate and lose their surface area after absorbing moisture. The advanced superabsorbent sheets, on the other hand, are prepared by

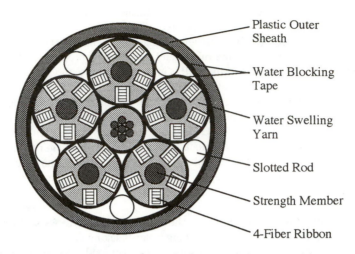

Fig. 7 Cross sectional image of a optical communication cable.

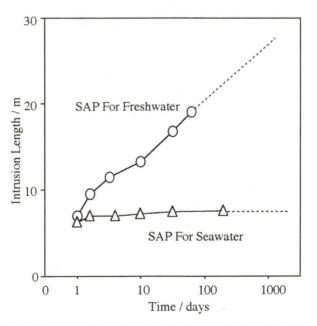

Fig. 8 Seawater intrusion length in a optical cable with a water blocking tape.

polymerizing hydrophilic monomers in or on non-woven fabrics (SAP-incorporate sheet) (*11-14*), or by carboxylating cellulosic non-woven fabrics (*15*).

The preparation process of SAP-incorporate sheets basically consists of two steps. First, hydrophilic monomers are sprayed on or soaked into a non-woven fabric, and second, the monomers are polymerized by heat, radiation, or radical initiators.

The features of the SAP-incorporate sheets are the following:

(1) SAP immobility on or in the substrate
(2) processability such as heat-sealing and cutting
(3) ease of handling
(4) applicability to vertical surfaces and ceilings
(5) high moisture absorbing and releasing rate.

The high moisture absorbing and releasing rate is based on the immobility of the SAP in the sheet. There will be no surface area loss in the SAP region by agglomeration, unlike the laminate sheet. Table III shows basic performance of SAP incorporated sheets. These sheets are very useful as dew prevention materials and as dewatering filters for lubricating oils.

Figure 9 shows the moisture absorbing and releasing performance of a SAP incorporated sheet. First, the SAP sheet absorbed moisture at 90 % relative humidity until it reached the equilibrium absorption. When the humidity decreased to 60 %, the sheet released the absorbed moisture and reached a new equilibrium. The sheet responded to the cyclic humidity change for a long period of time without any loss in its performance. This is very beneficial in moisture control, and when the SAP sheet is applied on walls or ceilings, it can keep a room under a desired humidity.

Figure 10 shows the moisture-controlling property of a SAP incorporated sheet in a closed container. In this experiment, the web was pre-soaked to a particular water content and then put in a closed container, and the humidity change in the container is investigated. The SAP incorporated sheet effectively controls the humidity in the container compared to the control, and the humidity depends on the pre-soaked water content.

SAP for Agriculture and Horticulture. SAP for agriculture and horticulture is one of the most important application because SAPs could transform deserts into green fertile lands. The SAP powder for hygiene applications is about 300 μm in diameter. However, for agriculture and horticulture, it is more suitable that the particles have a larger diameter (ca. 1-3 mm) and higher gel strength because small or soft gels fill spaces in soil and then prevent roots from breathing or water from draining. For the same reasons, the amount of SAPs mixed in soil is also very important, and about 0.1 wt% of SAP is usually added to soil.

Figure 11 shows the influence of the amount of a poly(acrylic acid)-type SAP designed for agricultural applications on water retention in soil. In this experiment, the moisture content in soil was measured after initial watering without additional watering during the experiment. As the amount of the SAP mixed in the soil increases, the soil can retain more moisture for longer times, and the plants live longer after germination. Without the SAP, the soil dries up very quickly and the seed never germinates.

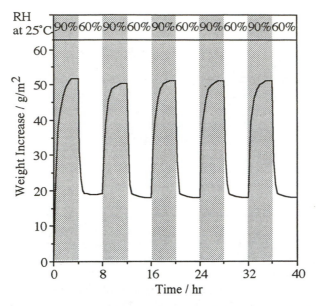

Fig. 9 Moisture absorbing/releasing performance of the SAP-incorporate sheet. (Adapted from ref. 3)

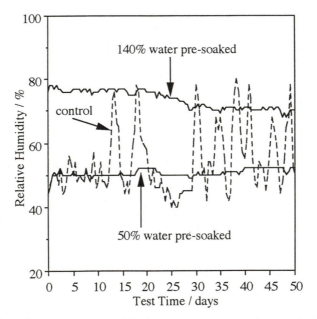

Fig. 10 Moisture control in a closed container by the SAP-incorporate sheet.

Table III. Performance of Web-type SAP

		Type A	Type B
Absorbency	Deionized Water	5000 g/m^2	800 g/m^2
		75 g/g	12 g/g
	Saline Solution	1000 g/m^2	10 g/m^2
		17 g/g	500 g/g
Moisture Absorption			
(at 25 °C, 90% RH)		45 g/m^2	50 g/m^2
Thickness		0.2 mm	0.2 mm
Basis Weight		70 g/m^2	70 g/m^2

In addition to the water retention, the SAP also improves the air content in soil. As the result, the following advantages can be expected by the SAP addition in soil:

(1) longer interval between watering
(2) less watering
(3) lower death rate of plants
(4) higher growing rate of plants
(5) higher fertilizer retention.

Less watering decreases the accumulation of salts contained in water or soil, and this may greatly help in the greening of deserts.

Figure 12 shows the yields of Brassica Rapa, a kind of Chinese vegetable commonly known as "Komatsuna," up to the third harvest in the same pots using the water-saving condition. In this experiment, 0.15% of SAP powder was mixed in the soil with fertilizers, and Komatsuna was planted and cultured in pots for 30 days. After harvesting, new Komatsuna was planted and cultured again in the same pot. The whole procedure was repeated up to the third harvest, and Komatsuna harvested each time was weighed. The poly(acrylic acid)-type SAP increased the yield about 10% compared to the control. The improvements are even greater when sulfonate type SAP is used. This is because the sulfonate SAP is less sensitive to polyvalent cations from soil, water, or fertilizers and can maintain its performance for a long time.

Figure 13 is a picture of a greenhouse in the Egyptian desert where a water-saving culture method with SAP has been investigated. The Ministry of International Trade and Industry of Japan and SAP Manufacturers in Japan have been co-operating with the Egyptian government on this project since 1990. This project, named "the Green Earth project," is one of the R&D support projects in the Japanese official development assistance program. In the greenhouses, vegetables and fruits were cultured, and SAP dramatically improved the crop under the water-saving condition. Although there are still some problems in SAP cost or gel stability against light, SAP will be practically used for agriculture in deserts or greening of deserts in the near future.

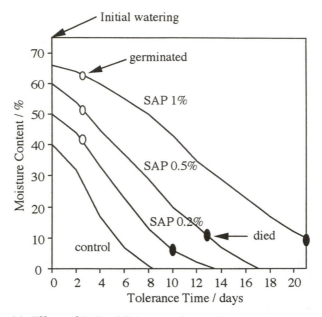

Fig. 11 Effect of SAP addition on the moisture content in soil.

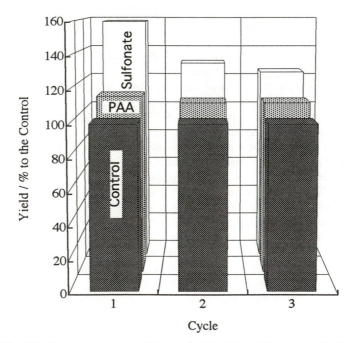

Fig. 12 Persistence test of the poly(acrylic acid)-type and the sulfonate-type SAPs as a water retainer.

Other Applications

Debris Flow Control. The Public Works Research Institute of the Ministry of Construction of Japan has been directing a project to investigate the effect of SAP addition to reduce the damage caused by debris flow (*13*). A debris flow is a strong, destructive mud flow that contains great many big rocks. It occurs in swift streams or in valleys during or after an extremely heavy rainfall. Once it reaches residential areas, it causes serious disaster.

Figure 14 shows the schematic image of the test facility to simulate a debris flow. The narrow gutter simulates a valley and the wide area at the bottom of the facility simulates an alluvial fan. Soil and stones are placed at the top of the gutter and swept away by tremendous amount of water. SAP is added below the gate in an amount of 1.4-12.5 % of the soil. When SAP is added to the debris flow, the SAP thickens the debris flow and prevents it from reaching deeply into residential areas. Thus, the SAP will dramatically reduce the damage by debris flow. Current issues on this application are the method of SAP addition to the real debris flow and the removal of the mud containing SAP gel.

Artificial Snow. The first indoor ski slope in the world was opened near Tokyo in 1991. The ski slope is 50 meters wide and 120 meters long. In order to prepare the artificial snow from a SAP, the SAP is swollen 100-120 times by water, spread a thickness of about 15 cm on the slope, and frozen by the cooling system. Then, the frozen gel is crushed to a particular size and its surface groomed to make the feeling of the gel closer to real snow (*17*). By using SAP gels as the artificial snow, the temperature in the building can be as high as 15 °C. This is much higher than that needed when the artificial snow is made of crushed ice or made with a snow gun (the temperature must be held below 5 °C). Close investigation of the artificial snow has been made by comparing with real snow, and the feeling of the artificial snow is very close to "powder snow."

Gel Actuators. Gel actuators are one of the most active research areas since they are the key material for artificial muscles that will make it possible to build a robot whose movement is human-like.

Gel actuators are prepared by adapting the absorbency dependence on the swelling condition (*14*) or the phase transition of a gel (*15*). In both cases, various swelling conditions can be used to control the actuator such as temperature, light, electric field, and change of the liquid. The gel as an actuator requires quick response to the change of conditions, high force generation, and high gel stability during repeated swelling and shrinking. Kurauchi et al., have used an electric field to bend an ionic polymer gel, and they have made a "Gel fish", whose fin is made of a gel, swim and have constructed a "Gel hand" with gel fingers grab an egg softly without breaking the egg (*20*).

Conclusion

Use of poly(acrylic acid)-type SAPs has spread widely and quickly for hygiene applications because of their high absorbency, high safety, and low cost. However,

Fig. 13 A greenhouse in Egypt for investigation of SAP performance as a water retainer.

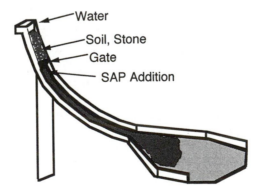

Fig. 14 A debris flow test facility.

they are not always suitable for other applications such as agriculture, horticulture and construction since they are so especially designed for hygiene applications. Some of the problems could not be solved simply by optimizing their size, shapes or crosslink density.

Years of SAP research have resulted in great advances in our understanding of gels. New, high performance SAPs have been developed to better fit with particular applications. The new high performance SAPs cover very wide application areas from water-swelling rubber to the control of debris flow. Their superior performance in each application has been demonstrated. There may still be many unknown SAP properties. As they are unveiled by further research, new high performance superabsorbent polymers will be developed in new forms or with new functions to broaden the areas of application.

Literature Cited

1. Weaver, M. O.; Bagley, E. B.; Fanta, G. F.; Doane, W. M. US 3,981,100 (1976).
2. Flory, P. J. *Principles of Polymer Chemistry*; Cornell University Press, Ithaca, NY, 1953, p. 584.
3. Shimomura, T. *Hyomen,* **1991**, 29, 495
4. Kimura, K.; Hatsuda, T.; Nagasuna, K. US 5,164,459 (1992).
5. Tanaka, H. *Sci. Am.,* **1981**, 249, 124
6. Tsubakimoto, T.; Shimomura, T.; Kobayashi, H. JP 62149335 (1987).
7. Tsubakimoto, T.; Shimomura, T.; Kobayashi, H. JP 62149336 (1987).
8. Kobayashi, H.; Okamura, K.; Sano, Y.; Shimomura, T. WO 8810001 (1988).
9. Shimomura, T.; Kobayashi, H.; Miyake, K.; Okamura K. JP 1129087 (1989).
10. Sheu, J. J. US 5,163,115, (1992).
11. Ericksen, P. H.; Nguyen, H. V.; Oczkowski, B.; Olejnik, T. A. EP 40087 (1981).
12. Ito, K.; Shibano, T. JP 62053309 (1987).
13. Miyake, K.; Harada, N.; Kimura, K.; Shimomura, T. EP 370646 (1990).
14. Suzuki, T. JP 60151381 (1985).
15. Nakahara, Y.; Egami, A.; Kataobe, A. JP 61089364 (1986).
16. Mizuyama, T.; Kurihara, J.; Suzuki, H. *Technical Memorandum of Public Works Research Institute,* **1989**, No. 277913
17. Morioka; K., Nakahigashi; S. *Refrigeration,* **1992**, 67, 28
18. Shiga, T.; Hirose, Y.; Okada, A.; Kurauchi, T. *J. Appl. Polym. Sci.,* **1993**, 47, 113
19. E. Sato, E.; Tanaka, T. *J. Chem. Phys.,* **1988**, 89, 1695
20. Kurauchi, N. *Kagaku to Kyoiku,* **1991**, 39, 618

RECEIVED March 25, 1994

Chapter 10

Water-Blocking, Optical-Fiber Cable System Employing Water-Absorbent Materials

Kazuo Hogari and Fumihiro Ashiya

NTT Telecommunication Field Systems Research and Development Center, Tokai-mura, Ibaraki-ken 319–11, Japan

A new optical fiber cable system, which consists of high-density and high-fiber-count water-blocking optical fiber cables and water sensors employing water absorbent materials, was developed for optical transmission networks. This system is capable of suppressing water penetration in a cable and detecting water penetration into a cable joining point. These cables are maintenance-free in term of water penetration, because it can be limited to a short length in the cable. Water sensors, which are installed on a monitoring fiber ribbon at each cable joining point, can detect water penetration at that point.

Optical fiber networks are expected to be capable of providing high-bit rate digital and broad-band analog services to subscribers. Optical fiber cables and fiber joining technologies are important if optical fiber networks are to be economically and efficiently constructed. In addition, efficient maintenance systems are necessary for optical fiber networks, because optical fibers can experience strength degradation when exposed to a variety of stresses, such as bending and tensile stress. This strength degradation can be accelerated if the fiber is exposed to water over long period of time.

In order to prevent water penetration, various types of optical fiber cable, such as gas-pressurized cables[1], jelly-filled cables and compound-filled cables[2],[3], have been developed and used. Although they have stable transmission and mechanical characteristics, they have several shortcomings. Though gas-pressurized cable is perfectly capable of preventing water penetration, maintenance is expensive in comparison with other cables. Although it is easy to apply jelly- or compound-filled cables to the optical fiber network to suppress water penetration, they are difficult to handle during manufacturing and fiber joining.

In a cable joining point, fibers are bent and accommodated after they have been spliced together or connected. Bent fibers suffer greater stress than those in cables, and the fiber strength degradation is accelerated if water penetrates a cable joining point. Thus, it is necessary to detect water penetration using the sensor and remove any water in a cable joining point.

Recently, we have developed a new water-blocking optical fiber cable system composed of high-density and high-fiber-count water-blocking optical fiber cables and water sensors which employ water absorbent materials. This paper describes the structures and characteristics of these cables and water sensors.

0097–6156/94/0573–0128$08.00/0
© 1994 American Chemical Society

Water-Blocking Cable System

The outline of the system is shown in Fig.1. This system is composed of high-density and high-fiber-count water-blocking optical cables and water sensors. The water-blocking cables are maintenance-free, because they restrict water penetration in the cables to a short length. The water sensors, which are installed on a monitoring fiber ribbon at each cable joining point, are used to detect water penetration at that point.

Water-Blocking Cable

Cable Structure. The structure of optical fiber ribbon, which has the advantages of the compactness, ease of mass-splicing[4] and mass manufacturing, is shown in Fig.2. Each optical fiber was coated with a primary coating.   Four or eight coated fibers were assembled and arranged linearly, and coated with a common ribbon coating. The structures of the high-density and high-fiber-count water-blocking cable (WB cable) are shown in Fig.3. Four types of cable were developed: 100, 300, 600 and 1000 fiber cable. The 100 and 300 fiber cables are composed of 4-fiber ribbons, a plastic slotted-rod, a strength member, copper wires, water-blocking tapes and a polyethylene (PE) sheath. Five 4-ribbons are accommodated tightly in slots and stacked closely at the bottom of each slot[1]. The copper wires are inserted in one slot. The water-blocking tape is wrapped around the slotted-rod. With the 600 and 1000 fiber cables, each plastic rod has 5 slots, and five 4-fiber ribbons or 8-fiber ribbons are inserted tightly into each slot. The slotted-rod units are stranded along a strength member. Water-blocking tapes are wrapped around each slotted-rod unit and the stranded rods as shown in Fig.4.

The water-blocking tape is shown in Fig.5. The water-blocking tape is composed of a base tape and absorbent powder[crosslinked poly (sodium acrylate)]. The powder is fixed to the surface of the base tape with adhesive, and remains in place during cable manufacturing and laying. When the sheath is cut and water comes into contact with the water-blocking tape, the absorbent powder is released and spreads into the empty spaces in the cable. The absorbent powder absorbs water, swells and become a gel. The gelled powder forms a dam to suppress the penetration of water into the cable.

These cables have advantages in terms of handling, manufacturing and weight, compared with jelly-filled and compound-filled optical fiber cables.

Cable Characteristics. The WB cables described in the previous section were manufactured. The mode field diameter and effective cut-off wavelength of the single mode fiber were 8.5 μm~10.5 μm and 1.10 μm~1.29 μm, respectively. All the coating materials were ultraviolet curable resins. The primary coated fiber diameter was 0.25 mm. The 4-fiber ribbon was 1.1 mm wide and 0.4 mm thick, and the 8-fiber ribbon was 2.1 mm wide and 0.4 mm thick. The transmission and mechanical characteristics for water free cables were measured and the results are shown in Table 1. It is found that WB cables have stable transmission and mechanical characteristics.

The water penetration length into the WB cables was measured. The water penetration test method is shown in Fig.6. The height of the water is 1 m and the solution is artificial sea-water. The stripped sheath length is 25 mm. A test result for 1000 fiber WB cable is shown in Fig.7. The penetration length was 0.8 m after a day and only 0.9 m after 90 days. From these results, we estimate that the water would penetrate only about 1 m in 20 years (7300 days).

In addition, the loss changes were measured at high and low temperatures in the water-absorbed WB cable. Water was poured into a WB cable along 10 m using a pump. The results are shown in Fig.8. No loss changes over measurement accuracy

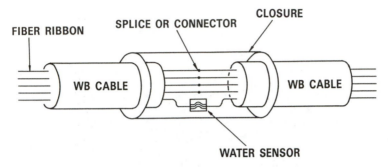

Figure 1 Water-blocking cable system.

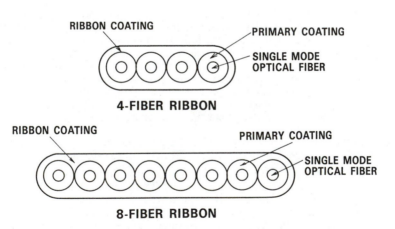

Figure 2 Fiber ribbon structures.

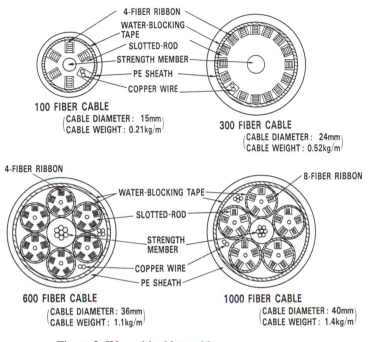

Figure 3 Water-blocking cable structures.

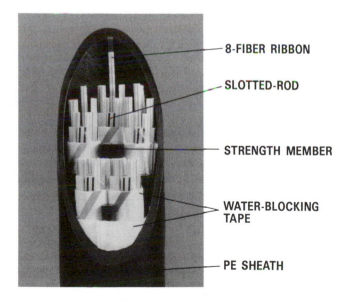

Figure 4 1000-fiber water-blocking cable.

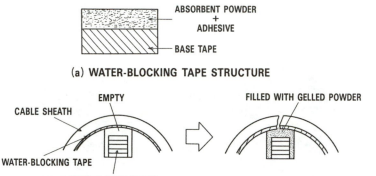

(a) WATER-BLOCKING TAPE STRUCTURE

(b) ABSORBENT POWDER IN CABLE

Figure 5 Water-blocking tape.

Table 1   TRANSMISSION AND MECHANICAL CHARACTERISTICS

(TYPICAL VALUES : NO PENETRATION)

ITEM	100 FIBER CABLE	300 FIBER CABLE	600 FIBER CABLE	1000 FIBER CABLE
LOSS : MEASUREMENT WAVELENGTH $\lambda=1.3\mu m$	0.36 dB/km	0.36 dB/km	0.35 dB/km	0.37 dB/km
TENSION : 2000N (100 FIBER CABLE, 300 FIBER CABLE) 8000N (600 FIBER CABLE, 1000 FIBER CABLE) [$\lambda=1.55\mu m$, CABLE LENGTH 10m]	0.02 dB	0.02 dB	0.03 dB	0.02 dB
LATERAL FORCE : 20N/mm [$\lambda=1.55\mu m$, CABLE LENGTH 500mm]	0.02 dB	0.02 dB	0.01 dB	0.01 dB
BENDING : BENDING RADIUS 6D [$\lambda=1.55\mu m$, D : CABLE DIAMETER]	0.03 dB	0.02 dB	0.03 dB	0.02 dB
SQUEEZING : 2000N, 250mmR (100 FIBER CABLE, 300 FIBER CABLE) 8000N, 600mmR (600 FIBER CABLE, 1000 FIBER CABLE) [$\lambda=1.55\mu m$, CABLE LENGTH 50m]	0.02 dB	0.02 dB	0.02 dB	0.04 dB
TEMPERATURE : -40˚C~60˚C [$\lambda=1.55\mu m$]	0.04 dB/km	0.05 dB/km	0.05 dB/km	0.03 dB/km

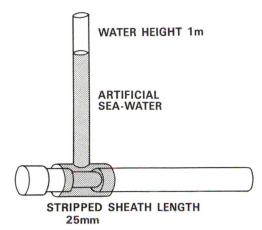

Figure 6 Water penetration test method.

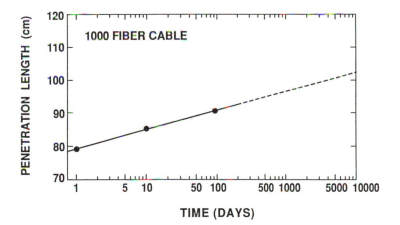

Figure 7 Result of water penetration test.

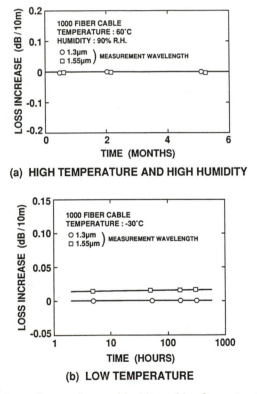

(a) HIGH TEMPERATURE AND HIGH HUMIDITY

(b) LOW TEMPERATURE

Figure 8 Loss change of water-blocking cable after water absorption.

can be detected at 1.3 μm and 1.55 μm wavelengths.

A 650 m length of the 100 fiber WB cable was installed in a duct in our laboratory. The sheath and water-blocking tape of this cable were removed at an arbitrary point and water was continuously introduced into the cable from this point. The optical loss change was measured and the result is shown in Fig.9. The loss changes were found to be small for 1000 days. It is therefore confirmed that WB cables have stable characteristics even if water is absorbed.

Fiber Strength of Water-Blocking Cable. These WB cables have stable transmission and mechanical characteristics as described in the previous section. However, some absorbed water remains in a cable after penetration and this may lead to a reduction in fiber strength. Therefore, the failure probability of a fiber is estimated using eq. (1) assuming that fiber strength in a water-absorbed WB cable would decrease to same degree as in water[5].

$$F=1-\exp\{N_pL[1-(1+(\varepsilon_r/\varepsilon_p)^n(t_r/t_p))^{(m/(n-2))}]\}, \tag{1}$$

where F is the failure probability of a fiber, L is fiber length, N_p is the failure number during a proof test, ε_r is fiber strain, ε_p is the applied fiber strain during the proof test, t_r is the time in use, t_p is the proof test time, m is the constant to determine the Weibull distribution, and n is the fatigue coefficient. The calculated result is shown in Fig.10. When water penetrates 1 km into a cable, the increase in the failure probability is more than 50 times that estimated for water free cable. On the other hand, when water penetrates less than 3 m into the cable, the increase in the failure probability is very small. With WB cables, water penetration can be limited to less than 3 m as described in the previous section, and the decrease of fiber strength is very small from this result.

From these results, it is found that the WB cables have stable transmission, mechanical and water-blocking characteristics for optical fiber networks, and it is not necessary to monitor and detect water penetration in the cable.

Water Sensor

At cable joining points, fibers are bent and accommodated after they have been spliced together or connected, and the bending strain causes a failure probability increase. When water penetrates the cable joining point, the failure probability is further increased. The failure probability of a fiber in a closure is also estimated using eq. (1). The calculated result is shown in Fig.11. When water penetrates a cable joining point, the increase in the failure probability is more than 10 times when there is no penetration. On the other hand, if the water penetration can be detected in a short time, the increase in the failure probability is small. Therefore, it is necessary to monitor and detect such water penetration, and remove the water. We have developed a water sensor in order to detect water penetration easily using an optical time domain reflectometer (OTDR).

The water sensor structure is shown in Fig.12. The water sensor consists of water absorbent material and a bender. When water penetrates into the cable joining point, the absorbent material absorbs the water, swells and pushes up the bender. The bender bends the monitoring fiber ribbon, and the optical loss in the bent fiber increases. The water sensor was manufactured with the dimensions of 40×4×20 mm. The sensor before and after water penetration is shown in Fig.13. The loss increase was measured at 24 water sensors after water penetration and the result is shown in Fig.14. It was found that the volume of absorbent material increased with loss increases of more than 2 dB after 1 hour.

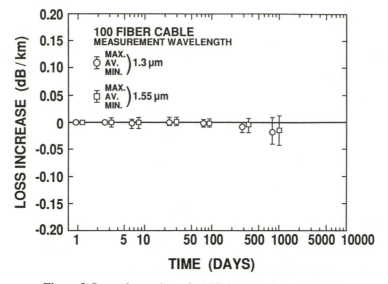

Figure 9 Loss change in an installed water-blocking cable.

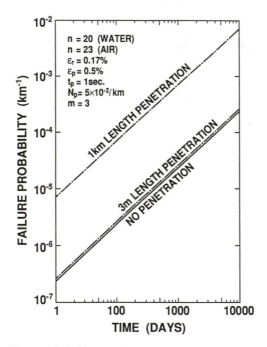

Figure 10 Failure probability of a fiber in a cable.

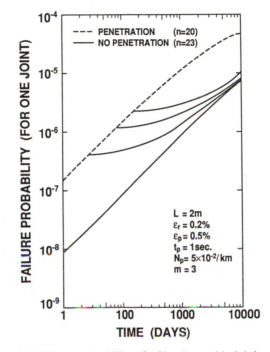

Figure 11 Failure probability of a fiber in a cable joining point.

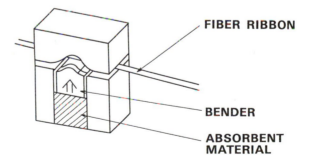

Figure 12 Water sensor structure.

(a) BEFORE WATER PENETRATION

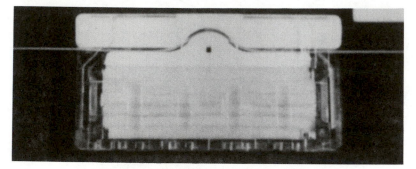

(b) AFTER WATER PENETRATION

Figure 13 Water sensor before and after water penetration.

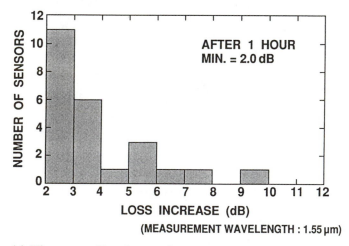

Figure 14 Histogram of loss increase in water sensors after water penetration.

The monitoring system at a cable joining point is shown in Fig.15. When there is no water in the cable joining point, the sensor does not operate and no loss increase is observed. On the other hand, once the water enters the cable joining point, the sensor operates and optical loss of the monitoring fiber increases. The optical loss increase can be measured using an OTDR. A sensor was installed in each cable joining point along the optical fiber network. The OTDR waveforms before and after water penetration into the cable joining point are shown in Fig.16. It is found that water penetration can easily be detected from the optical loss increase.

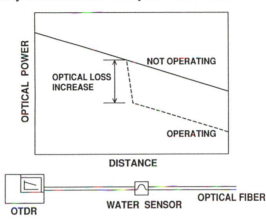

Figure 15 Detection of water penetration in cable joining point.

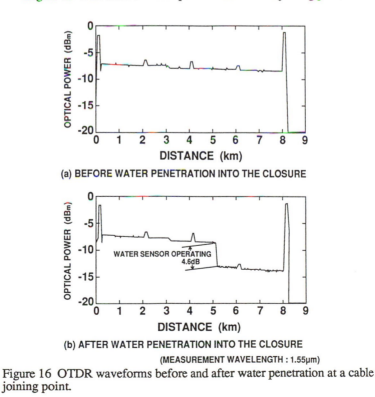

(a) BEFORE WATER PENETRATION INTO THE CLOSURE

(b) AFTER WATER PENETRATION INTO THE CLOSURE

(MEASUREMENT WAVELENGTH : 1.55μm)

Figure 16 OTDR waveforms before and after water penetration at a cable joining point.

Conclusion

A new optical fiber cable system, which consists of high-density and high-fiber-count water-blocking optical fiber cables and water sensors employing water absorbent materials, was developed. The water-blocking cables have the advantages of ease of manufacture, ease of handling, light weight and stable transmission, mechanical and water-blocking characteristics, and they are maintenance-free. The water sensor can detect water penetration at a cable joining point. This system is capable of maintaining optical fiber cables without difficulty. This system has already been introduced into optical fiber networks.

Acknowledgment

The authors thank Ishihara K., Kawase M. and Morimitsu T. for their support and encouragement.

References
(1) Kawase, M.; Fuchigami, T.; Matsumoto, M.; Nagasawa, S.; Tomita, S.; Takashima, S.; J. Lightwave Tech. **1989**, vol.LT-7, pp.1675-1681.
(2) Kameo, Y.; Horima, H.; Tanaka, S.; Ishida, Y.; Koyamada, Y.; Proc. of 30th International Wire & Cable Symposium **1981**, pp.236-243.
(3) Mitchell, D. M.; Sabia, R.; Proc. of 29th International Wire & Cable Symposium **1980**, pp.15-25.
(4) Tachikura, M.; Kashima, N.; J. Lightwave Tech. **1984**, vol.LT-2, pp.25-31.
(5) Mitsunaga, Y.; Katsuyama, Y.; Kobayashi, H.; Ishida, Y.; J. Appl. Phys. **1982**, vol.53, pp.4847-4853.

RECEIVED March 25, 1994

INDEXES

Author Index

Affiliation Index

Subject Index

Production: Charlotte McNaughton
Indexing: Deborah H. Steiner
Acquisition: Anne Wilson
Cover design: Neal Clodfelter

Printed and bound by Maple Press, York, PA

Highlights from ACS Books

For further information and a free catalog of ACS books, contact:
American Chemical Society
Distribution Office, Department 225
1155 16th Street, NW, Washington, DC 20036
Telephone 800–227–5558

Bestsellers from ACS Books

The ACS Style Guide: A Manual for Authors and Editors
Edited by Janet S. Dodd
264 pp; clothbound ISBN 0–8412–0917–0; paperback ISBN 0–8412–0943–X

The Basics of Technical Communicating
By B. Edward Cain
ACS Professional Reference Book; 198 pp;
clothbound ISBN 0–8412–1451–4; paperback ISBN 0–8412–1452–2

Chemical Activities (student and teacher editions)
By Christie L. Borgford and Lee R. Summerlin
330 pp; spiralbound ISBN 0–8412–1417–4; teacher ed. ISBN 0–8412–1416–6

Chemical Demonstrations: A Sourcebook for Teachers,
Volumes 1 and 2, Second Edition
Volume 1 by Lee R. Summerlin and James L. Ealy, Jr.;
Vol. 1, 198 pp; spiralbound ISBN 0–8412–1481–6;
Volume 2 by Lee R. Summerlin, Christie L. Borgford, and Julie B. Ealy
Vol. 2, 234 pp; spiralbound ISBN 0–8412–1535–9

Chemistry and Crime: From Sherlock Holmes to Today's Courtroom
Edited by Samuel M. Gerber
135 pp; clothbound ISBN 0–8412–0784–4; paperback ISBN 0–8412–0785–2

Writing the Laboratory Notebook
By Howard M. Kanare
145 pp; clothbound ISBN 0–8412–0906–5; paperback ISBN 0–8412–0933–2

Developing a Chemical Hygiene Plan
By Jay A. Young, Warren K. Kingsley, and George H. Wahl, Jr.
paperback ISBN 0–8412–1876–5

Introduction to Microwave Sample Preparation: Theory and Practice
Edited by H. M. Kingston and Lois B. Jassie
263 pp; clothbound ISBN 0–8412–1450–6

Principles of Environmental Sampling
Edited by Lawrence H. Keith
ACS Professional Reference Book; 458 pp;
clothbound ISBN 0–8412–1173–6; paperback ISBN 0–8412–1437–9

Biotechnology and Materials Science: Chemistry for the Future
Edited by Mary L. Good (Jacqueline K. Barton, Associate Editor)
135 pp; clothbound ISBN 0–8412–1472–7; paperback ISBN 0–8412–1473–5

For further information and a free catalog of ACS books, contact:
American Chemical Society
Distribution Office, Department 225
1155 16th Street, NW, Washington, DC 20036
Telephone 800–227–5558

L2238